Probability and Statistics Applications for Environmental Science

Probability and Statistics Applications for Environmental Science

Stacey J. Shaefer
Louis Theodore

CRC Press
Taylor & Francis Group
Boca Raton   London   New York

CRC Press is an imprint of the
Taylor & Francis Group, an informa business

CRC Press
Taylor & Francis Group
6000 Broken Sound Parkway NW, Suite 300
Boca Raton, FL 33487-2742

ISBN 13: 978-0-367-45316-9 (pbk)
ISBN 13: 978-0-8493-7561-3 (hbk)

Visit the Taylor & Francis Web site at
http://www.taylorandfrancis.com

and the CRC Press Web site at
http://www.crcpress.com

Library of Congress Cataloging-in-Publication Data

Shaefer, Stacey J.
    Probability and statistics for environmental science / Stacey J. Shaefer and Louis Theodore.
        p. cm.
    Includes bibliographical references and index.
    ISBN 0-8493-7561-4 (alk. paper)
    1. Environmental sciences--Statistical methods. I. Theodore, Louis. II. Title.

GE45.S73S46 2006
363.7001'5195--dc22
                                      2006050889

Dedication

To Tricia (S.S.),
my sister and best friend,
thanks for always being there
and
Frankie (L.T.),
who made this happen

Preface

As for a future life,
Every man must judge for himself
Between conflicting vague probabilities.

Charles Darwin: Life and Letters
Edited by Francis Darwin
1887

It is becoming more and more apparent that the education process must include courses that will develop the interest and needs of the student. It is no secret that the teaching of probability and statistics is now considered essential in most curricula and is generally accepted as one of the fundamental branches of mathematics. This course, or its equivalent, is now a required undergraduate course at most colleges and universities and is slowly and justifiably finding its way into other curricula.

The question now arises: Why another textbook on probability and statistics? A very reasonable question. After all, there are over 100 books in this area. The answer lies in the approach — taken by many of the authors' previous books — in that the textbook was written at a level that the beginning student could understand. In effect it was written for students, not colleagues and statisticians.

At this time, most of the texts in this field are considered by some to be too advanced for the undergraduate student. The present text is intended to overcome the difficulties and sometimes frightening experiences a beginning student encounters on being introduced to this subject. The authors' aim is to offer the reader the fundamentals of probability and statistics with appropriate practical applications and to serve as an introduction to the specialized and more sophisticated texts in this area.

The present book has primarily evolved from a Theodore Tutorial prepared by Frank Taylor titled "Probability and Statistics." Material has also been drawn by notes prepared by one of the authors (many years ago) for a one-semester three-credit course given to senior students at Manhattan College; the course was also offered as an elective to the other disciplines. It is assumed that the student has already taken basic courses in the sciences, and there should be a minimum background in mathematics through differential equations. The text roughly places equal emphasis on theory and applied mathematics. However, depending on the needs and desires of the reader, either the fundamentals or applications may be emphasized — and the material is presented in a manner to permit this.

In constructing this text from the aforementioned tutorial and notes, the problem of what to include and what to omit has been particularly difficult. Topics of interest to all technical individuals and students have been included. A general discussion of the philosophy and contents of the text follows.

The material is divided into three sections. Section I keys in on principles and fundamentals. Some of the topics here include (see Table of Contents) set notation, introduction to probability distributions, and the estimation of the mean and variance. Section II is primarily concerned with traditional statistics applications. Much of this material centers around the uses of probability distributions, including how they relate to reliability and failure theory. Subject matter includes many of the important distributions, Monte Carlo methods, and fault and event trees. Section III delves into what some have come to define as contemporary statistics. Topics covered here include hypothesis testing, Student's t-tests and Chi-square tests, regression analysis, analysis of variance (ANOVA), and nonparametric tests. An introduction to Design of Experiments (Part A) is included in the Appendix along with a Glossary of Terms (Part B) and a host of key tables (Part C).

The reader should note that a few of the examples are concerned with the game of dice, ordinarily not a topic of interest to a technical individual. However, one of the driving forces behind the writing of this book is to generate royalty income to help support one of the author's gambling habits. And, it turns out that dice is the favorite casino game of this author.

An important point needs to be made. There have been numerous occasions during one of the author's 47-year tenure as an educator when students solved a problem using a packaged program such as Excel, MathCAD, etc. For example, the problem could have involved the solution to a differential equation or the regression of some data. On being questioned how the packaged program performed the calculation, the student almost always responded with something to the effect of, "I don't know and I don't care." For this reason, the reader should note that no attempt was made to introduce and apply packaged computer programs that are presently available. The emphasis in this introductory text in probability and statistics was to provide the reader with an understanding of fundamental principles in order to learn how statistical methods can be used to obtain answers to questions within one's own subject matter specialty; becoming "computer literate" in this field was not the objective.

We are deeply indebted to the aforementioned Frank Taylor who in a very real sense allowed this text to become a reality. It should also be noted that significant material was drawn from lecture notes and problems prepared again for a probability and statistics course offered during the late 1960s and early 1970s; the source of some of this material is no longer known.

Stacey J. Shaefer
Louis Theodore
2007

Contents

SECTION II   *Traditional Applications*

APPENDIX A

APPENDIX B

APPENDIX C

Section I

Fundamentals

1 PAC: Permutations and Combinations

INTRODUCTION

The problem of calculating probabilities of *objects* or *events* in a finite group — defined as the *sample space* in the next chapter — in which equal probabilities are assigned to the elements in the sample space requires counting the elements which make up the events. The counting of such events is often greatly simplified by employing the rules for permutations and combinations.

Permutations and combinations deal with the grouping and arrangement of objects or events. By definition, each different ordering in a given manner or arrangement with regard to order of all or part of the objects is called a *permutation*. Alternately, each of the sets which can be made by using all or part of a given collection of objects without regard to order of the objects in the set is called a *combination*. Although, permutations or combinations can be obtained *with replacement* or *without replacement*, most analyses of permutations and combinations are based on sampling that is performed without replacement; i.e., each object or element can be used only once.

For each of the two *with/without* pairs (with/without regard to order and with/without replacement), four subsets of two may be drawn. These four are provided in Table 1.1.

Each of the four paired subsets in Table 1.1 is considered in the following text with accompanying examples based on the letters A, B, and C. To personalize this, the reader could consider the options (games of chance) one of the authors faces while on a one-day visit to a casino. The only three options normally considered are dice (often referred to as craps), blackjack (occasionally referred to as 21), and pari-mutual (horses, trotters, dogs, and jai alai) simulcasting betting. All three of these may be played during a visit, although playing two or only one is also an option. In addition, the order may vary and the option may be repeated. Some possibilities include the following:

Dice, blackjack, and then simulcast wagering
Blackjack, wagering, and dice
Wagering, dice, and wagering
Wagering and dice (the author's usual sequence)
Blackjack, blackjack (following a break), and dice

In order to simplify the examples that follow, dice, blackjack, and wagering are referred to as *objects*, represented by the letters A, B, and C, respectively.

TABLE 1.1
Subsets of Permutations and Combinations

Permutations (With Regard to Order)	Combinations (Without Regard to Order)
Without replacement	Without replacement
With replacement	With replacement

---

1. Consider a scenario that involves three separate objects, A, B, and C. The arrangement of these objects is called a *permutation*. There are six different orders or permutations of these three objects possible, as follows, while noting that ABC ≠ CBA.

$$\begin{array}{ccc} ABC & BAC & CAB \\ ACB & BCA & CBA \end{array}$$

Thus, BCA would represent blackjack, wagering, and dice.

The number of different permutations of n objects is always equal to $n!$, where $n!$ is normally referred to as factorial n. Factorial n or $n!$ is defined as the product of the n objects taken n at a time and denoted as $P(n, n)$. Thus,

$$P(n, n) = n! \tag{1.1}$$

With three objects, the number of permutations is $3! = 3 \times 2 \times 1 = 6$. Note that $0!$ is 1.

The number of different permutations of n objects taken r at a time is given by

$$P(n, r) = \frac{n!}{(n-r)!} \tag{1.2}$$

(**Note**: The permutation term P also appears in the literature as ^nP_r or $P(^n_r)$.) For the three objects A, B, and C, taken two at a time, ($n = 3$, and $r = 2$). Thus,

$$P(3, 2) = \frac{3!}{(3-2)!} = \frac{(3)(2))(1)}{1} = 6$$

These possible different orders, noting once again that AB ≠ BC, are

$$\begin{array}{ccc} AB & BC & AC \\ BA & CB & CA \end{array}$$

Consider now a scenario involving n objects in which these can be divided into j sets with the objects within each set being alike. If $r_1, r_2, ..., r_j$ represent the number of objects within each of the respective sets, with $n = r_1 + r_2 + ... + r_j$, then the number of permutations of the n objects is given by

$$P(n; r_1, r_2, ..., r_j) = \frac{n!}{r_1! r_2! ... r_j!} \tag{1.3}$$

This represents the number of permutations of n objects of which r_1 are alike, r_2 are alike, and so on. Consider, for example, 2 As, 1 B, and 1 C; the number of permutations of these 4 objects is

$$P(4; 2, 1, 1) = \frac{4!}{2!1!1!} = 12$$

The 12 permutations for this scenario are as follows:

| AABC | ABAC | ABCA | BAAC | BACA | BCAA |
| AACB | ACAB | ACBA | CAAB | CABA | CBAA |

2. Consider the arrangement of the same three objects in Example 1, but obtain the number of permutations (with regard to order) with replacement, PR. There are 27 different permutations possible.

AAA	BBB	CCC
AAB	BBA	CCB
AAC	BBC	CCA
ABA	BAB	CBC
ABB	BAA	CBB
ACA	BCB	CAC
ACC	BCC	CAA
ABC	BAC	CBA
ACB	BCA	CAB

For this scenario,

$$PR(n, n) = (n)^n \qquad (1.4)$$

so that

$$PR(3, 3) = (3)^3 = 27$$

For n objects taken r at a time,

$$PR(n, r) = (n)^R \qquad (1.5)$$

3. The number of different ways in which one can select r objects from a set of n without regard to order (i.e., the order does not count) and without replacement is defined as the number of *combinations*, C, of the n objects taken r at a time. The number of combinations of n objects taken r at a time is given by

$$C(n, r) = \frac{P(n,r)}{r!} = \frac{n!}{r!(n-r)!} \qquad (1.6)$$

(Note: The combination term C also appears in the literature as C_r^n or $C(_r^n)$.)

For the ABC example taken two letters at a time, one has

$$C(3,2) = \frac{3!}{2!1!}$$

$$= \frac{(3)(2)(1)}{(2)(1)(1)}$$

$$= 3$$

TABLE 1.2
Describing Equations for Permutations and Combinations

	Without Replacement	With Replacement	Type
With regard to order	$P(n,r) = \dfrac{n!}{(n-r)!}$ (Equation 1.2)	$PR(n,r) = (n)^r$ (Equation 1.5)	Permutation
Without regard to order	$C(n,r) = \dfrac{n!}{r!(n-r)!}$ (Equation 1.6)	$CR(n,r) = C(n,r) + (n)^{r-1}$ (Equation 1.7)	Combination

The number of combinations becomes

$$AB \qquad AC \qquad BC$$

Note that the combination BA is not included since AB = BA for combinations.

4. The arrangement of n objects, taken r at a time without regard to order and with replacement is denoted by *CR* and given by (personal notes: L. Theodore)

$$CR(n, r) = C(n, r) + (n)^{n-1} \tag{1.7}$$

with

$$CR(n, n) = C(n, n) + (n)^{n-1}; \quad r = n \tag{1.8}$$

For the ABC example taken two letters at a time, one employs Equation 1.7.

$$CR = C(3,2) + (3)^{2-1}$$
$$= 3 + 3$$
$$= 6$$

The number of combinations becomes

$$\begin{array}{ccc} AA & BB & CC \\ AB & AC & BC \end{array}$$

Table 1.1 may now be rewritten to include the describing equation for each of the above four subsets. This is presented in Table 1.2.

PROBLEMS AND SOLUTIONS

PAC 1: IDENTIFYING LABORATORY SAMPLES

Discuss the relative advantages or disadvantages of identifying a laboratory sample as follows:

1. With three numbers
2. With three letters

Solution

This is an example of sampling with regard to order and with replacement, because the numbers may be replaced (reused) or replicated. For this case, Equation 1.5 is employed.

1. Three numbers give the following solution:

$$PR(n,r) = (n)^r; \quad n = 10, r = 3$$

$$= (10)(10)(10) = 10^3$$

$$= 1,000$$

2. With three letters, the solution is as follows:

$$PR(26,3) = (26)(26)(26) = (26)^3$$

$$= 17,576$$

Obviously, the latter choice (three letters) provides greater flexibility, because numerous identification possibilities are available.

PAC 2: LICENSE PLATE SETS

Determine the number of possible license plates drawn from 26 letters and 10 integers that consist of 7 symbols. An example is the current license plate number of one of the authors: CCY9126. Note that this calculation again involves a situation in which the order of the symbol does matter and the symbol may be repeated.

Solution

Based on the problem statement there are 26 letters and 10 integers. Therefore, 36 symbols are available to choose from. The number of permutations of r symbols that can be drawn from a pool of n symbols is again given by

$$PR(n, r) = (n)^r$$

For this case, $r = 7$ and $n = 36$. The number is therefore

$$PR(n,r) = (36)^7$$

$$= 7.84 \times 10^{10}$$

Obviously, there are a rather large number of permutations.

PAC 3: LABORATORY LABELS

How many four-digit laboratory labels can be formed from the numbers 0 to 9 if a number cannot be repeated in the four-digit label.

Solution

These are permutations without replacement with $n = 10$ and $r = 4$. Therefore, Equation 1.2 applies:

$$P(n,r) = \frac{n!}{(n-r)!} = P(10,4)\frac{10!}{6!} = (10)(9)(8)(7)$$

$$= 5,040$$

Note once again that

$$n! = (n)(n-1) \ldots (2)(1)$$

and

$$0! = 1$$

PAC 4: CHEMICAL PERMUTATIONS

Determine the number of 4-element chemical compounds that can theoretically be generated from a pool of 112 elements. Assume each element counts only once in the chemical formula and that the order of the elements in the compound matters. An example of a three-element compound is H_2SO_4 (sulfuric acid) or CH_3OH (methanol). An example of a four-element compound is $NaHCO_3$.

Solution

As discussed earlier, each different ordering or arrangement of all or part of a number of symbols (or objects) in which the order matters is defined as a *permutation*. There are 112 elements to choose from in this application. For this problem, the describing equation once again is given by Equation 1.2.

$$P(n,\ r) = \frac{n!}{(n-r)!}$$

Based on the problem statement, $n = 112$ and $r = 4$. Therefore,

$$P(112,4) = \frac{112!}{(112-4)!} = \frac{112!}{108!}$$

$$= (112)(111)(110)(109)$$

$$= 1.49 \times 10^8$$

As with the previous problem, PAC 2, this too is a large number. This number would be further increased if the number of a particular element appearing in the chemical formula was greater than one, for example, HCN (hydrogen cyanide) vs. C_3H_3N (acrylonitrile). However, a more realistic scenario would involve a calculation in which the order does not matter. This is treated in Problem PAC 8.

Two points need to be made, one concerning the calculation and the other concerning the chemistry of the compound:

1. For calculations involving large factorials, it is often convenient to use an approximation known as Stirling's formula:

$$n! \approx (2\pi)^{0.5} e^{-n} n^{n+0.5}$$

2. For a real-world "viable" compound, the elements involved in this application must be capable of bonding.

PAC 5: WORD PERMUTATIONS

Find the number of four-letter words — with all four letters being different — that can be formed from the letters of the word BELMONT (the most beautiful of all racetracks and a home away from home during the summer months for one of the authors). Also determine the number of three-letter words that can be formed that contain only consonants.

Solution

Once again, this involves a permutation calculation without replacement. The number of four-letter words that can be formed from BELMONT is given by

$$P(7,4) = \frac{7!}{3!} = (7)(6)(5)(4)$$

$$= 840$$

Because there are five consonants, the number of three-letter words that can be formed is

$$P(5, 3) = \frac{5!}{2!} = (5)(4)(3) = 60$$

PAC 6: BASKETBALL TEAM STARTING FIVE

How many different starting basketball teams (of 5 players each) can be formed from a team consisting of 12 players?

Solution

This involves a combination calculation because the position on the team does not matter. For this application, the combination equation must be applied, i.e., Equation 1.6.

$$C(n, r) = \frac{n!}{r!(n-r)!}$$

Noting that $n = 12$ and $r = 5$,

$$C(12,5) = \frac{12!}{5!7!} = \frac{(12)(11)(10)(9)(8)}{(5)(4)(3)(2)}$$

$$= 792$$

PAC 7: LINES GENERATED FROM POINTS

Calculate the number of straight lines that can be drawn through six points. Assume no three points form a straight line.

Solution

The reader should note that this solution involves a combination (not a permutation) calculation with $n = 6$ and $r = 2$. Thus,

$$C(6,2) = \frac{6!}{2!4!} = \frac{(6)(5)}{(2)}$$

$$= 15$$

This calculation can be verified by performing the procedure by longhand, i.e., by placing six dots on a sheet of paper and drawing lines through the various combinations of points. For ease of analysis, it is suggested that points be located on the circumference of a circle.

PAC 8: COMBINATIONS OF CHEMICAL ELEMENTS

Refer to Problem PAC 4. Recalculate the number of chemical compounds that can be formed in which the arrangement of symbols is such that the order does not matter, as with a chemical compound. For example, HCN is the same as CNH.

Solution

For this case, the combination equation (Equation 1.6) applies:

$$C(n, r) = \frac{n!}{r!(n-r)!}$$

Substituting once again gives

$$C(112,4) = \frac{112!}{(4!)(112-4)!} = \frac{112!}{(4!)(108!)}$$

$$= \frac{(112)(111)(110)(109)}{(4)(3)(2)(1)}$$

$$= 6.21 \times 10^6$$

As expected, the result is lower than that generated in Problem PAC 4.

PAC 9: MULTIPLE PERMUTATIONS

Refer to Problem PAC 5. How many four-letter words can be formed from BELMONT according to the following rules?

1. The word begins and ends with a constant.
2. The word begins with the letter L and ends in a vowel.
3. The word begins with the letter T and also contains the letter L.

Solution

This involves multiple permutations and more than one calculation is involved. For this case, permutations apply because the order of the words (W) does matter, and the product of the permutations involved in the calculation gives the total number of words.

1. Here, two consonants are drawn from a pool of five consonants, leaving two letters to be drawn from a pool of five letters. Therefore,

$$W = P(5,2)P(5,2)$$
$$= (5 \times 4)(5 \times 4)$$
$$= 400$$

2. Once L has been selected as the first letter and a vowel as the last letter, the two middle letters are chosen from a pool of five letters. Noting that the vowel is drawn from a pool of two letters leads to

$$W = (1)P(5,2)P(2,1)$$
$$= (1)(5 \times 4)(2 \times 1)$$
$$= (1)(20)(2)$$
$$= 40$$

3. For this case, the remaining two letters need to be drawn from a pool of five letters. Therefore,

$$W = (1)P(5,2)P(3,1)$$
$$= (1)(5 \times 4)(3 \times 1)$$
$$= 60$$

PAC 10: MULTIPLE COMBINATIONS

Exposure to a nanoagent has four ways in which it can lead to a major health effect and twelve ways to a minor effect. How many ways may an individual be exposed to the following?

1. Two major and two minor health effects
2. One major and four minor health effects

Solution

Because the order in which the effects appear or impact on an individual does not matter, an analysis involving combinations without replacement is required. The number of combined health effects (HE) is thus,

1. For two major and two minor effects:

$$HE = C(4,2)C(12,2)$$

$$= \left(\frac{4!}{2!2!}\right)\left(\frac{12!}{10!2!}\right)$$

$$= \left(\frac{4 \times 3}{2}\right)\left(\frac{12 \times 11}{2}\right)$$

$$= 396$$

2. For one major and four minor effects:

$$HE = C(4,1)C(12,4)$$

$$= \left(\frac{4!}{3!1!}\right)\left(\frac{12!}{8!4!}\right)$$

$$= (4)\left(\frac{12 \times 11 \times 10 \times 9}{4 \times 3 \times 2}\right)$$

$$= 1,980$$

PAC 11: NEW YORK STATE LOTTO

The wife of one of the authors requested that the spouse calculate the odds of picking the six correct numbers in a LOTTO drawing. The New York State LOTTO selects from a pool of 59 numbers (1 to 59).

Solution

This requires a traditional combination calculation because the drawing is performed without replacement and without regard to order. Equation 1.6 therefore applies with $n = 59$ and $r = 6$.

$$C(n,r) = \frac{n!}{r!(n-r)!} = \frac{59!}{6!53!}$$

$$= \frac{(59)(58)(57)(56)(55)(54)}{(6)(5)(4)(3)(2)(1)}$$

$$= \frac{3.244 \times 10^{10}}{720}$$

$$= 45,060,000$$

The odds are therefore approximately 45 million to one.

The reader should note that $1 buys two chances for LOTTO.

PAC 12: NEW YORK STATE MEGA MILLIONS LOTTERY

Refer to Problem PAC 11. The wife has also asked for the odds with the New York State Mega Millions lottery. This lottery requires the selection of one correct number from a pool of numbers from 1 to 46, which is then followed by the selection of 5 numbers from 1 to 56.

Solution

As will be discussed in later chapters in this section, this probability calculation involves two *mutually exclusive* events. The result is obtained from the product of the probability of each event. Thus,

$$P = P(5)P(1)$$

where

$$P(5) = C(56, 5)$$

and

$$P(1) = C(46, 1)$$

Substituting yields

$$P(5) = C(56,5) = \frac{56!}{51!5!}$$

$$= \frac{(56)(55)(54)(53)(52)}{(5)(4)(3)(2)(1)}$$

$$= \frac{4.587 \times 10^8}{120}$$

$$= 3,820,000$$

and

$$P(1) = C(46, 1) = 46$$

Therefore,

$$P = (3,820,000)(46)$$

$$= 175,700,000$$

The Mega Millions probability is approximately 175 million to 1. As with LOTTO, prizes are provided for selecting less than six numbers; for example, three numbers win $7.

2 PDI. Probability Definitions and Interpretations

INTRODUCTION

Probabilities are nonnegative numbers associated with the outcomes of so-called random experiments. A random experiment is an experiment whose outcome is uncertain. Examples include throwing a pair of dice, tossing a coin, counting the number of defectives in a sample from a lot of manufactured items, and observing the time to failure of a tube in a heat exchanger, or a seal in a pump, or a bus section in an electrostatic precipitator. The set of possible outcomes of a random experiment is called the *sample space* and is usually designated by S. Then $P(A)$, the probability of an event A, is the sum of the probabilities assigned to the outcomes constituting the subset A of the sample space S. Consider, for example, tossing a coin twice. The sample space can be described as

$$S = \{HH, HT, TH, TT\}$$

If probability 1/4 is assigned to each element of S, and A is the event of at least one head, then

$$A = \{HH, HT, TH\}$$

The sum of the probabilities assigned to the elements of A is 3/4. Therefore, $P(A) = 3/4$.

The description of the sample space is not unique. The sample space S in the case of tossing a coin twice could be described in terms of the number of heads obtained. Then

$$S = \{0, 1, 2\}$$

Suppose probabilities 1/4, 1/2, and 1/4 are assigned to the outcomes 0, 1, and 2, respectively. Then A, the event of at least one head, would have for its probability,

$$P(A) = P\{1, 2\} = 3/4$$

How probabilities are assigned to the elements of the sample space depends on the desired interpretation of the probability of an event. Thus, $P(A)$ can be interpreted as *theoretical relative frequency*, i.e., a number about which the relative frequency of event A tends to cluster as n, the number of times the random experiment is performed, increases indefinitely. This is the objective interpretation of probability. Under this interpretation, to say that $P(A)$ is 3/4 in the aforementioned example means that if a coin is tossed n times, the proportion of times one or more heads occurs clusters about 3/4 as n increases indefinitely.

As another example, consider a single valve that can stick in an open (O) or closed (C) position. The sample space can be described as follows:

$$S = \{O, C\}$$

Suppose that the valve sticks twice as often in the open position as it does in the closed position. Under the theoretical relative frequency interpretation, the probability assigned to element O in S would be 2/3, twice the probability assigned to the element C. If two such valves are observed, the sample space S can be described as

$$S = \{OO, OC, CO, CC\}$$

Assuming that the two valves operate independently, a reasonable assignment of probabilities to the elements of S, as just listed, should be 4/9, 2/9, 2/9, and 1/9. The reason for this assignment will become clear after consideration of the concept of independence in the problem section of Chapter 5. If A is the event of at least one valve sticking in the closed position, then

$$A = \{OC, CO, CC\}$$

The sum of the probabilities assigned to the elements of A is 5/9. Therefore, $P(A) = 5/9$.

Probability $P(A)$ can also be interpreted subjectively as a measure of degree of belief, on a scale from 0 to 1, that the event A occurs. This interpretation is frequently used in ordinary conversation. For example, if someone says, "The probability that I will go to a casino tonight is 90%," then 90% is a measure of the person's belief that he or she will go to a casino. This interpretation is also used when, in the absence of concrete data needed to estimate an unknown probability on the basis of observed relative frequency, the personal opinion of an expert is sought. For example, an expert might be asked to estimate the probability that the seals in a newly designed pump will leak at high pressures. The estimate would be based on the expert's familiarity with the history of pumps of similar design. The definition of probability is revisited in Problem PDI. 3.

PROBLEMS AND SOLUTIONS

PDI 1: SAMPLES AND POPULATIONS

Discuss the difference between a sample and a population.

Solution

When there is a set of n observations, one may wish to use the values of these observations in many instances to estimate certain characteristics of a larger number of observations that have not been made. In other words, the interest is not directly on a particular set of observations, but rather in using them to estimate something about a potentially larger set. One refers to the observations as a *sample*. The larger "supply," which one may have, is called a *population*. From the characteristics of a sample of observations, one can estimate similar characteristics for the population.

PDI 2: STATISTIC VS. PARAMETER

Discuss the difference between a statistic and a parameter.

Solution

When one calculates some measured value of a sample, it is referred to as a *statistic*. A statistic is any characteristic of a set of observations calculated from a sample. The measure corresponding to a statistic obtained from a population is referred to as a *parameter*. In most cases of interest, the values of parameters are unknown and must be estimated by the statistics derived from a sample. Thus, the sample value is an estimate of the population value.

PDI 3: PROBABILITY DEFINITIONS

The physical interpretation of probability is a subject of some controversy. Provide three alternative interpretations of probability.

Solution

The three major interpretations of probability are as follows:

1. The *classical* (or *a priori*) interpretation: If an event can occur in N equally likely and mutually exclusive ways and if n of these ways have an outcome or attribute E, then the *probability* of occurrence of E is n/N. The term *mutually exclusive* has several definitions, one of which is that each event has nothing in common. See Problem BTH 1 in Chapter 4 for additional details.
   The drawback of this interpretation is that it is not clear when certain events are "equally likely" and it does not indicate what happens when events are not equally likely. This has led to the second interpretation.
2. The *frequency* (or *objective*) interpretation: If an experiment is conducted N times, and a particular outcome or event E occurs n times, then the limit of n/N as N tends to infinity is called the *probability* of event E.
   This is the most popular present-day interpretation. However, a small but growing school of modern statisticians advocate the third definition.
3. The *subjectivist* (or *personalistic*) interpretation: *Probability* represents the *degree of belief* one holds in a particular proposition, as was discussed earlier.
   This is a much more liberal and all-inclusive interpretation. For example, the subjectivist considers it quite reasonable to speak of "the probability that Lou Theodore will be the next mayor of New York City" or the "probability that this book will sell." These statements have no technical meaning to the classical statistician or to the frequentist.

Despite the controversy on the physical interpretation of *probability*, the axiomatic definitions provided in this chapter and Chapter 3 are accepted.

PDI 4: SAMPLE SPACE

Four environmental management design projects have been ranked A, B, C, and D in the order of their economic attractiveness, with A denoting the best and D, the worst. Two of these designs end up going forward to receive a detailed economic analysis. Determine the probability of the following:

1. The two selected are the two most competent designs.
2. They are the least competent designs.

Solution

This is an example of sampling without replacement in which the order does not matter (without regard to order). The possible choices are

$$(A, B), (A, C), (A, D), (B, C), (B, D), \text{ and } (C, D)$$

This can be verified by application of Equation 1.6 where $n = 2$, $r = 2$:

$$C(n,r) = C(2,2) = \frac{4!}{2!2!}$$

$$= \frac{24}{4}$$

$$= 6$$

These six pairs of elements from the designs constitute the sample space, S. Because the selection of each pair is random, event 1 has one pair (A, B), so the probability is 1/6. Event 2 has the same probability because only (C, D) is the choice.

PDI 5: DICE SAMPLE SPACE

A pair of fair dice is rolled once. Fair dice are defined as not loaded, i.e., each side occurs with equal relative frequency. Describe the sample space, S.

Solution

Note that the description of S is not unique. Choose a description appropriate to the problem under consideration. For this case,

$$S = \{2, 3, 4, 5, 6, 7, 8, 9, 10, 11, 12\}$$

PDI 6: DICE PROBABILITY

A pair of fair dice is rolled once. Calculate the probability of obtaining a sum greater than 8.

Solution

Assign probabilities to the elements of S on the basis of relative frequency. The sum 2 can occur in only one way: 1 on the first die and 1 on the second die. The sum 3 can occur in two ways: 1 on the first and 2 on the second, 2 on the first and 1 on the second. Therefore, the long-run relative frequency of the sum 3 should be twice that of the sum 2, i.e., if x is the probability assigned to 2, the probability assigned to 3 should be $2x$. Similar comparison of the way each sum can be obtained with the way in which the sum 2 is obtained yields probability assignments to the elements of S, as shown in Table 2.1.

Because $36x$, the total sum of the probabilities assigned, must equal 1, $x = 1/36$. The probabilities assigned to the elements of S are shown in Table 2.2.

Represent the event in the problem statement whose probability is to be calculated as a subset of S and include in the subset all of the elements of S that correspond to the occurrence of the event:

$$\text{Event of sum} > 8 = \{9, 10, 11, 12\}$$

Obtain the probability of the event as the sum of the probabilities assigned to the elements of the subset representing the event. Use Table 2.2 for the assigned probabilities.

$$P(\text{sum} > 8) = 4/36 + 3/36 + 2/36 + 1/36$$

$$= 10/36 = 5/18$$

TABLE 2.1
Dice Sample Space

Sum	Probability
2	x
3	$2x$
4	$3x$
5	$4x$
6	$5x$
7	$6x$
8	$5x$
9	$4x$
10	$3x$
11	$2x$
12	x

TABLE 2.2
Dice Probabilities

Sum	Probability (for each)
2, 12	1/36
3, 11	2/36
4, 10	3/36
5, 9	4/36
6, 8	5/36
7	6/36

PDI 7: DICE SYMMETRY I

If a pair of fair dice is rolled once, calculate the probability of obtaining a sum less than 6.

Solution

For this application,

$$\text{Event of sum} < 6 = \{2, 3, 4, 5\}$$

$$= 1/36 + 2/36 + 3/36 + 4/36$$

$$= 10/36 = 5/18$$

This is the same result obtained in Problem PDI 6. Note that the result could have been obtained directly from Problem PDI 6 because of the "symmetry" of the sample space.

PDI 8: DICE SYMMETRY II

If a pair of fair dice is rolled once, calculate the probability of obtaining a number greater than 5.

Solution

For this application, one notes that the probability of obtaining a number greater than 5 is given by 1 less than the probability of obtaining a number less than 6. (The sum of the probabilities assigned must again equal 1). Thus, from Problem PDI 7,

$$\text{Event of sum} > 5 = \{6, 7, 8, 9, 10, 11, 12\}$$

$$= 5/36 + 6/36 + 5/36 + 4/36 + 3/36 + 2/36 + 1/36$$

$$= 26/36 = 13/18$$

3 SNE. Set Notation

INTRODUCTION

Various combinations of the occurrence of any two events A and B can be indicated in set notation as shown in Figure 3.1. Note that the hatched areas represent the symbol at the bottom of each figure.

Venn diagrams provide a pictorial representation of these events. In set terminology $\bar{A}$ is called the *complement* of A. The notation AB is called the *intersection* of A and B, i.e., the set of elements in both A and B. An alternate notation in the literature for the intersection of A and B is $A \cap B$. The notation $A + B$ is called the *union* of A and B, i.e., the set of elements in A, B, or both A and B. An alternate notation for the union of A and B is $A \cup B$.

When events A and B have no elements in common they are said to be *mutually exclusive*. A set having no elements is called the *null set* and is designated by ϕ. Thus, if events A and B are *mutually exclusive* then $AB = \phi$. As can be seen in Figure 3.1.b, the union of A and B consists of three *mutually exclusive* events: $A\bar{B}$, AB, and $\bar{A}B$.

The algebra of sets — Boolean algebra — governs the way in which sets can be manipulated to form equivalent sets. The principal Boolean algebra laws used for this purpose are as follows:

Commutative laws

$$A + B = B + A \tag{3.1}$$

$$AB = BA \tag{3.2}$$

Associative laws

$$(A + B) + C = A + (B + C) = A + B + C \tag{3.3}$$

$$(AB)C = A(BC) = ABC \tag{3.4}$$

Distributive laws

$$A(B + C) = AB + BC \tag{3.5}$$

$$A + BC = (A + B)(A + C) \tag{3.6}$$

Absorption laws

$$A + A = A \tag{3.7}$$

$$AA = A \tag{3.8}$$

DeMorgan's laws

$$\overline{AB} = \bar{A} + \bar{B} \tag{3.9}$$

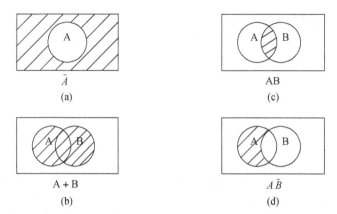

FIGURE 3.1 Venn diagrams: (a) $\bar{A}$ = A does not occur (similarly $\bar{B}$ = B does not occur), (b) A + B = A occurs or B occurs (*or* used in mutually inclusive sense to indicate the occurrence of A, B, or both A and B), (c) AB = A occurs and B occurs, (d) A$\bar{B}$ = A occurs and B does not occur.

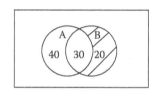

FIGURE 3.2 Venn diagram temperature readings: $\bar{A}B$ (the shaded area) is equal to 20.

If numbers replace the letter symbols for sets, the commutative and associative laws become familiar laws of arithmetic. In Boolean algebra, the first of the two distributive laws has an analogous counterpart in arithmetic; the second does not. In some health and hazard risk analysis applications, Boolean algebra is used to simplify expressions for complicated events.

PROBLEMS AND SOLUTIONS

SNE 1: Venn Diagram – Temperature Readings

Of the 120 temperature readings recorded at a hazardous waste incinerator, 70 were recorded in °F. A total of the 40 readings in °F were below the expected or typical average. However, a total of 50 readings were above the average. How many readings were both not recorded in °F and above the average? Draw a Venn diagram.

Solution

Denote A to represent the readings in °F and B to represent the known readings above the average. This is represented in Figure 3.2.

SNE 2: Coin Flipping Sample Space

A coin is flipped. If heads occur, the coin is flipped again; otherwise, a die is tossed once. List the elements in the sample space S.

Solution

For this case, the sample space is

$$S = \{(H, H)\ (H, T)\ (T, 1)\ (T, 2)\ (T, 3)\ (T, 4)\ (T, 5)\ (T, 6)\}$$

SNE 3: COIN FLIPPING PROBABILITY

Refer to Problem SNE 2. If A is the event of tails on the first flip of the coin and B is the event of a number less than 4 on the die, find $P(A)$, $P(B)$, $P(\bar{B})$, $P(A\bar{B})$, and $P(A + B)$.

Solution

Note that the probability of tossing a head (or a tail) is 1/2 and the probability of a number appearing on a die is 1/6. Refer to the sample space in the previous problem. For this case, one may assign probability 1/4 to (H, H) and 1/4 to (H, T), and assign probability 1/12 to each of the other 6 elements of S.

$$P(A) = \{(T, 1)\ (T, 2)\ (T, 3)\ (T, 4)\ (T, 5)\ (T, 6)\}$$

$$= \{(1/12)(1/12)(1/12)(1/12)(1/12)(1/12)\}$$

$$= 6(1/12) = 1/2 \text{ (as expected)}$$

$$P(B) = \{(T, 1)\ (T, 2)\ (T, 3)\}$$

$$= (1/6)(1/6)(1/6)$$

$$= 3(1/6) = 1/2$$

$$P(\bar{B}) = 1 - P(B) = 1/2$$

The reader is referred to Equation 4.4 in Chapter 4 for additional details.

$$P(A\bar{B}) = \{(T, 4)\ (T, 5)\ (T, 6)\}$$

$$= \{(1/12)(1/12)(1/12)\}$$

$$= 3(1/12) = 1/4$$

The reader is referred to Equation 4.6 in Chapter 4 for additional details.

$$P(A + B) = P(A) + P(B) - P(AB)$$

Since

$$P(AB) = P(A)(B) = 1/2 - 1/4 = 1/4 = 1/4$$

Then

$$P(A + B) = 1/2 + 1/4 - 1/4$$

$$= 1/2$$

	ENG	SCI
NANO	54	12
NON-NANO	46	58

FIGURE 3.3 Technical/technology split.

SNE 4: University Poll

A recent poll of 170 incoming university students at a local university consisted of 100 students who chose to major in engineering, the remaining selecting a science program as their major. Of this total, 66 indicated a preference for the inclusion of environmental implications of nanotechnology courses in the curriculum.

1. How many students opted against nanomaterial?
2. If 46 engineering majors did not prefer the inclusion of nanotechnology courses, how many science students preferred the inclusion of nanomaterial?

Solution

1. As 66 of the 170 students preferred nanomaterial, 104 (170 – 66) did not want it.
2. As 100 students chose engineering, 70 students selected science. Because 104 are non-nanotechnology proponents and 46 of these are engineering majors, 58, i.e., 104 – 46 of the 104 are science majors. Therefore, only 12, i.e., 70 – 58 science students are not interested in nanotechnology. A modified-Venn "box" diagram of this scenario is provided in Figure 3.3.

SNE 5: National Science Foundation Poll

There are 50 proposals to be reviewed by a National Science Foundation (NSF) panel. The classification of the proposals is as follows: 15 fundamental science (*FS*), 20 applied science (*AS*), and 35 environmental management (*EM*) proposals. There were 20 *EM* proposals categorized as neither *AS* nor *FS*. In addition, 3 of the proposals received all three classifications, i.e., *FS*, *AS*, and *EM*, and 2 were classified as both *FS* and *AS*. How many proposals not in the *AS* category but in the *FS* category were also classified as *EM*?

Solution

The solution is provided in the Venn diagram in Figure 3.4. From the figure, $(\overline{AS})(FS)(EM) = 6$ The reader is left the exercise of verifying the results.

SNE 6: Nanotechnology Start-Up Company

There are 60 employees in a recent nanotechnology start-up company. The employees have the following degrees: environmental engineering (*E*), 25; environmental science (*S*), 33; and, business (*B*), 30. The number of employees who have both *E* and *S* degrees totals 15, 6 have all three degrees, 10 *B*s have neither *E* nor *S* degrees, and those who have *S* and *B* but not *E* number 10.

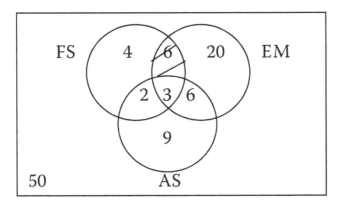

FIGURE 3.4 Venn diagram of National Science Foundation poll.

1. How many employees have *E*s but neither *S*s nor *B*s?
2. How many employees do not have a degree?

Solution

Refer to the Venn diagram in Figure 3.5. The Venn notation, gives a graphical view of the scenario.

1. In Venn notation, ⬚ :

$$E(\overline{S + B}) = 6$$

2. In Venn notation, ⬚ :

$$\overline{E + S + B} = 7$$

SNE 7: ENGINEERING STUDENTS I

In a group of 100 engineering students, 15 are studying to be chemical engineers, 70 are undergraduates, 10 of whom are studying to be chemical engineers. A student is selected at random from the group. Let *C* be the event that the student selected is studying to be a chemical engineer. Let *U* be the event that the student selected is an undergraduate. Find $P(\overline{C})$ and $P(C + U)$.

Solution

For this problem,

$$P(C) = 15/100$$
$$P(U) = 70/100$$
$$\text{or}$$
$$P(C) = 15/100; \quad P(U) = 70/100$$

$$P(CU) = 10/100$$

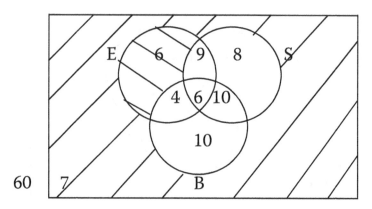

FIGURE 3.5 Venn diagram for the nanotechnology company.

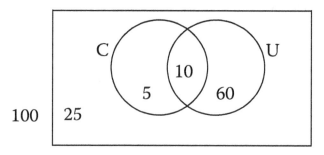

FIGURE 3.6 Venn diagram for engineering students.

Finally,

$$P(\bar{C}) = 1 - P(C) = 85/100$$

$$= 0.85$$

and

$$P(C + U) = P(C) + P(U) - P(CU)$$

$$= (15/100) + (70/100) - (10/100)$$

$$= 75/100$$

$$= 0.75$$

The reader may refer to Figure 3.6 in Problem SNE 8 for additional explanation.

SNE 8: ENGINEERING STUDENTS II

Refer to Problem SNE 7 and obtain $P(\bar{C}U)$, $P(\overline{CU})$, and $P(\bar{C}+\bar{U})$.

Solution

Based on set notation (see Figure 3.6),

$$P(\bar{C}U) = P(U) - P(CU)$$

$$= (70/100) - (10/100)$$

$$= 60/100$$

$$= 0.60$$

$$P(\overline{CU}) = P(\bar{C} + \bar{U}) = 1 - P(C + U)$$

$$= 1 - (75/100)$$

$$= 25/100$$

$$= 0.25$$

$$P(\bar{C} + \bar{U}) = P(\bar{C}) + P(\bar{U}) - P(\overline{CU})$$

$$= (85/100) + (30/100) - (25/100)$$

$$= 90/100$$

$$= 0.90$$

4 BTH. Basic Theorems

INTRODUCTION

The mathematical properties of $P(A)$, the probability of event A, are deduced from the following postulates governing the assignment of probabilities to the elements of a sample space, S.

1. $P(S) = 1$ (4.1)

2. $P(A) \geq 0$ for any event A   (4.2)

3. If A_1, ..., A_n, ... are *mutually exclusive*, then

$$P(A_1 + A_2 + \cdots + A_n + \cdots) = P(A_1) + P(A_2) + \cdots + P(A_n) + \cdots \tag{4.3}$$

In the case of a discrete sample space (i.e., a sample space consisting of a finite number or countable infinitude of elements), these postulates require that the numbers assigned as probabilities to the elements of S be nonnegative and have a sum equal to 1. These requirements do not result in complete specification of the numbers assigned as probabilities. The desired interpretation of probability must also be considered, as indicated in Chapter 2. However, the mathematical properties of the probability of any event are the same regardless of how this probability is interpreted. These properties are formulated in theorems logically deduced from the postulates mentioned earlier without the need for appeal to interpretation. The following are three basic theorems:

1. $P(\bar{A}) = 1 - P(A)$; Theorem 1   (4.4)

2. $0 \leq P(A) \leq 1$; Theorem 2   (4.5)

3. $P(A + B) = P(A) + P(B) - P(AB)$; Theorem 3   (4.6)

Theorem 1 says that the probability that A does not occur is one minus the probability that A occurs. Theorem 2 says that the probability of any event lies between 0 and 1. Theorem 3, the addition theorem, provides an alternative way of calculating the probability of the union of two events as the sum of their probabilities minus the probability of their intersection. The addition theorem can be extended to three or more events. In the case of three events A, B, and C, the addition theorem becomes

$$P(A + B + C) = P(A) + P(B) + P(C) - P(AB) - P(AC) - P(BC) + P(ABC) \tag{4.7}$$

For four events A, B, C, and D, the addition theorem becomes

$$P(A + B + C + D) = P(A) + P(B) + P(C) + P(D) - P(AB) - P(AC)$$

$$P(AD) - P(BC) - P(BD) - P(CD) + P(ABC) + P(ABD) + P(BCD)$$

$$+ P(ACD) - P(ABCD) \tag{4.8}$$

To illustrate the application of the three basic theorems, consider what happens when one draws a card at random from a deck of 52 cards. The sample space S may be described in terms of 52 elements, each corresponding to one of the cards in the deck. Assuming that each of the 52 possible outcomes would occur with equal relative frequency in the long run leads to the assignment of equal probability, 1/52, to each of the elements of S. Let A be the event of drawing a king and B the event of drawing a club. Thus, A is a subset consisting of four elements, each of which has been assigned probability 1/52, and $P(A)$ is the sum of these probabilities: 4/52. Similarly the following two probabilities are obtained:

$$P(B) = \frac{13}{52}$$

$$P(AB) = \frac{1}{52}$$

Application of Theorem 1 gives

$$P(\overline{A}) = \frac{48}{52}$$

$$P(\overline{B}) = \frac{39}{52}$$

Application of Theorem 3 gives

$$P(A + B) = \frac{4}{52} + \frac{13}{52} - \frac{1}{52} = \frac{16}{52}$$

The term $P(A + B)$, the probability of drawing a king or a club, could have been calculated without using the addition theorem by calling $A + B$, the union of A and B, a set consisting of 16 elements. (One obtains 16 by adding the number of kings, 4, to the number of clubs, 13, and subtracting the card that is counted twice — once as a king and once as a club.) Because each of the 16 elements in $A + B$ has been assigned probability 1/52, $P(A + B)$ is the sum of the probabilities assigned, namely, $(4 + 13 - 1)/52$, or 16/52.

PROBLEMS AND SOLUTIONS

BTH 1: MUTUALLY EXCLUSIVE EVENTS

Define mutually exclusive events.

Solution

If two or more events are such that not more than one of them can occur in a single trial, they are said to be *mutually exclusive*. For example, in a single draw from a deck of playing cards, the two events of obtaining a king and obtaining a queen are mutually exclusive. However the two events of drawing a queen and drawing a diamond are not mutually exclusive because in a particular draw a card may be both a queen and a diamond.

If $P_1, P_2, P_3, \ldots, P_n$ are the separate probabilities of the occurrence of mutually exclusive events, the probability P that some one of these events will occur in a single trial is

$$P = P_1 + P_2 + P_3 + \ldots + P_n \qquad (4.9)$$

This has been referred to as the probability addition rule.

BTH 2: Mutually Exclusive Application

Consider a deck of 52 cards. Find the probability of the following:

1. Drawing a picture card
2. Drawing a high card

(**Note:** A picture card is defined as the jack, queen, or king. A high card is normally defined as a 10, jack, queen, king, or ace.)

Solution

1. The probability of drawing a picture card is given by

$$P(\text{picture card}) = \frac{4}{52} + \frac{4}{52} + \frac{4}{52}$$

$$= \frac{1}{13} + \frac{1}{13} + \frac{1}{13}$$

$$= \frac{3}{13} = 0.231$$

2. The probability of drawing a high card is given by

$$P(\text{high card}) = \frac{1}{13} + \frac{1}{13} + \frac{1}{13} + \frac{1}{13} + \frac{1}{13}$$

$$= \frac{5}{13} = 0.385$$

BTH 3: Independent vs. Dependent Events

Describe the difference between independent and dependent events.

Solution

Two or more events are said to be *independent* if the probability of occurrence of any one of them
, is not influenced by the occurrence of any other; otherwise, the events are *dependent*. For example,
if cards are drawn in succession from a deck of cards without replacement, the appearance of a
king on the first draw and the appearance of a queen on the second draw depends on what happened
on the first draw. If after the first draw the card is replaced, then the outcome of the second draw
is independent of the result of the first draw. The repeated throwing of two dice in a "craps" game
are independent events because the outcome of any throw is not influenced by the outcome of the
other throws.

If P_1, P_2, P_3, ..., P_n are the individual probabilities of the occurrence of independent events,
the probability P that all these events will occur in a single trial is given by

$$P = P_1 P_2 P_3 \cdots P_n \tag{4.10}$$

This equation may be rephrased as follows: If two or more events are independent, the probability that they will all occur is the product of their separate probabilities. This can be referred to
as the *probability multiplication rule*.

BTH 4: Straight Passes

What is the probability of making five straight passes in a dice game on the first roll in the
following ways?

1. By rolling a 7
2. By rolling a 7 or 11

(**Note:** A pass is defined here as winning on the first roll. One can win on the first roll by
throwing either a 7 or an 11.)

Solution

1. Refer to Problem PDI 6 in Chapter 2. The probability of rolling a 7 is 6/36 or 1/6. As
   each pass is an independent event, Equation 4.10 applies:

$$P(5 \text{ passes}) = P_1 P_2 P_3 P_4 P_5; \quad P_i = \frac{1}{6}$$

Therefore,

$$P(5 \text{ passes}) = \left(\frac{1}{6}\right)\left(\frac{1}{6}\right)\left(\frac{1}{6}\right)\left(\frac{1}{6}\right)\left(\frac{1}{6}\right) = \left(\frac{1}{6}\right)^5$$

$$= \frac{1}{7776} = 1.286 \times 10^{-4}$$

2. From Problem PDI 6,

$$P(7 \text{ or } 11) = \frac{6}{36} + \frac{2}{36} = \frac{8}{36} = \frac{2}{9}$$

Therefore,

$$P(5 \text{ passes}) = P_1 P_2 P_3 P_4 P_5$$

$$= \left(\frac{2}{9}\right)\left(\frac{2}{9}\right)\left(\frac{2}{9}\right)\left(\frac{2}{9}\right)\left(\frac{2}{9}\right) = \left(\frac{2}{9}\right)^5$$

$$= \frac{32}{59,049} = 5.02 \times 10^{-4}$$

BTH 5: DEFECTIVE ITEMS

From a list of 100 items, 10 of which are defective, 2 items are drawn in succession. Find the probability that both items are defective if drawn as follows:

1. The first item is replaced before the second is drawn.
2. The first item is not replaced before the second is drawn.

Solution

1. Let $P(A)$ be the probability that the first item is defective and $P(B)$, the probability that the second item is defective. As the item is replaced and each drawing is an independent event, apply Equation 4.10.

$$P(1) = P(A) \, P(B) = \left(\frac{10}{100}\right)\left(\frac{10}{100}\right) = \frac{1}{100}$$

$$= 0.01$$

2. Because the item is not replaced (without replacement), the probability associated with the second drawing is affected by the first drawing. For this case,

$$P(2) = P(A) \, P(B) = \left(\frac{10}{100}\right)\left(\frac{9}{99}\right) = \frac{90}{9900} = \frac{1}{110}$$

$$= 0.0909$$

This case involves what is defined as *conditional probability* and is revisited in Chapter 5.

BTH 6: DEFECTIVE LIGHTBULBS

Three defective lightbulbs are inadvertently mixed with seven good ones. If two bulbs are chosen at random for a particle-sizing microscope, what is the probability that both are good?

Solution

Let $P(A)$ be the probability of a good bulb on the first drawing and $P(B)$, the probability of a good bulb on the second drawing. As with Problem BTH 5, the process involves drawing bulbs without replacement. Therefore,

$$P(2) = P(A)\ P(B) = \left(\frac{7}{10}\right)\left(\frac{6}{9}\right) = \frac{42}{90} = \frac{7}{15}$$

$$= 0.467$$

This problem, as with part 2 of the previous problem will be revisited in Chapter 5.

BTH 7: GENERATOR SWITCH FAILURE

The failure of a generator or a switch causes interruption of electrical power. If the probability of a generator failure is 0.02, the probability of a switch failure is 0.01, and the probability of both failing is 0.0002, what is the probability of electrical power interruption?

Solution

Label the events representing the causes of electrical power interruption; these events are generator failure and switch failure. Let A be the event of generator failure and B, the event of switch failure. Express the probability required in terms of a probability concerning A and B. Use set notation for the events.

$$P(\text{electrical power interruption}) = P(A \text{ or } B) = P(A + B)$$

From the data given obtain $P(A)$, $P(B)$, and $P(AB)$. Note that $P(AB)$ is the probability that both A and B occur.

$$P(A) = 0.02$$

$$P(B) = 0.01$$

$$P(AB) = 0.0002$$

Apply the addition theorem — Equation 4.6 — to find $P(A + B)$.

$$P(A + B) = P(A) + P(B) - P(AB)$$

$$= 0.02 + 0.01 - 0.0002$$

$$= 0.0298$$

Thus, the probability of electrical power interruption is 0.0298. Note that if the events are mutually exclusive, $P(AB) = 0$.

BTH 8: HEAT EXCHANGER MIX-UP

Six single-pass heat exchangers are inadvertently mixed with five double-pass exchangers, and it is not possible to tell them apart from appearance. If two exchangers are selected in succession, calculate the probability that one of them is a single-pass and the other a double-pass type.

Solution

This involves the following two mutually exclusive events without replacement:

1. Selecting a single-pass and then a double-pass exchanger
2. Selecting a double-pass and then a single-pass exchanger

Let $P(S)$ be the probability of drawing a single-pass exchanger and $P(D)$, the probability of drawing a double-pass exchanger. For case 1, the probability of these two events is

$$P(1) = P(S)\ P(D) = \left(\frac{6}{11}\right)\left(\frac{5}{10}\right) = \frac{30}{110} = \frac{3}{11}$$

$$= 0.273$$

Similarly, for case 2, the probability is

$$P(2) = P(S)\ P(D) = \left(\frac{5}{11}\right)\left(\frac{6}{10}\right) = \frac{30}{110} = \frac{3}{11}$$

$$= 0.273$$

As 1 and 2 are independent events, apply Equation 4.9,

$$P = P(1) + P(2)$$

$$= \frac{3}{11} + \frac{3}{11} = \frac{6}{11}$$

$$= 0.545$$

Note that the first result could have been obtained using the combination formula provided in Equation 1.6. Using this approach,

$$P(1) = \frac{\text{(number of ways of drawing an }S\text{)(number of ways of drawing a }D\text{)}}{\text{(total number of ways of drawing two exchangers)}}$$

$$P(1) = \frac{C(6,1)C(5,1)}{C(11,2)}$$

$$= \frac{(6)(5)}{(11)(10)} = \frac{30}{110} = \frac{3}{11}$$

5 COP. Conditional Probability

INTRODUCTION

The conditional probability of event B given A is denoted by $P(B|A)$ and defined as

$$P(B\backslash A) = P(AB)/P(A) \qquad (5.1)$$

where $P(B|A)$ can be interpreted as the proportion of B occurrences that first require the occurrence of A and $P(AB)$ is the proportion of both A and B occurrences (without regard to order). For example, consider the random experiment of drawing two cards in succession from a deck of 52 cards. Suppose the cards are drawn without replacement (i.e., the first card drawn is not replaced before the second is drawn). Let A denote the event that the first card is an ace and B the event that the second card is an ace. The sample space S can be described as a set of 52 times 51 pairs of cards. Assuming that each of these (52)(51) pairs has the same theoretical relative frequency, assign probability 1/(52)(51) to each pair. The number of pairs featuring an ace as the first and second card is (4)(3). Therefore,

$$P(AB) = \frac{(4)(3)}{(52)(51)}$$

$$= 0.0045$$

The number of pairs featuring an ace as the first card and one of the other 51 cards as the second is (4)(51). Therefore,

$$P(A) = \frac{(4)(51)}{(52)(51)}$$

$$= 0.0769$$

Applying the definition of conditional probability Equation 5.1 yields

$$P(B\backslash A) = P(AB)/P(A) = \frac{3}{51}$$

$$= \frac{0.0045}{0.0769}$$

$$= 0.0588$$

as the conditional probability that the second card is an ace, given that the first is an ace. The same result could have been obtained by computing $P(B)$ on the new sample space consisting of 51 cards, three of which are aces. This illustrates the two methods for calculating a conditional probability in terms of probabilities computed on the original sample space by means of the definition mentioned earlier. The second method uses the given event to construct a new sample space on which the conditional probability is computed.

Conditional probability also can be used to formulate a definition for the independence of two events, A and B. Event B is defined to be independent of event A if and only if

$$P(B\backslash A) = P(B) \qquad (5.2)$$

Similarly, event A is defined to be independent of event B if and only if

$$P(A\backslash B) = P(A) \qquad (5.3)$$

To illustrate the concept of independence, again consider the random experiment of drawing two cards in succession from a deck of 52 cards; this time, suppose that the cards are drawn with replacement (i.e., the first card is replaced in the deck before the second card is drawn). As before, let A denote the event that the first card is an ace, and B the event that the second card is an ace. Then

$$P(B\backslash A) = 4/52$$

and because $P(B|A) = P(B)$, B and A are independent events.

From the definition of $P(B|A)$ and $P(A|B)$, one can deduce the following modified form of the multiplication theorem, given in Equation 5.1.

$$P(AB) = P(A)\,P(B\backslash A) \qquad (5.4)$$

$$P(AB) = P(B)\,P(A\backslash B) \qquad (5.5)$$

The multiplication theorem provides an alternate method for calculating the probability of the intersection of two events. This is illustrated in Problem COP 2.

The multiplication theorem can also be extended to the case of three or more events. For three events A, B, and C, the multiplication theorem states

$$P(ABC) = P(A)\,P(B\backslash A)\,P(C\backslash AB) \qquad (5.6)$$

For four events A, B, C, and D, the multiplication theorem states

$$P(ABCD) = P(A)\,P(B\backslash A)\,P(C\backslash AB)\,P(D\backslash ABC) \qquad (5.7)$$

PROBLEMS AND SOLUTIONS

COP 1: Defective Transistors

A box of 100 transistors contains 5 defectives. A sample of two transistors is drawn at random from the box. What is the probability that both are defective in the following cases?

1. The sample is drawn without replacement.
2. The sample is drawn with replacement.

Solution

Let A be the event that the first transistor drawn is defective, and B the event that the second transistor drawn is defective. If the sample is drawn with replacement, the composition of the box remains the same after the first drawing. If the sample is drawn without replacement, then after the first transistor is drawn, there are 99 transistors in the box, 4 of which are defective. Therefore,

$$P(A) = 5/100$$

$$P(B\backslash A) = 5/100 \text{ if the sample is drawn } \textit{with replacement}$$

$$P(B\backslash A) = 4/99 \text{ if the sample is drawn } \textit{without replacement}$$

Use the multiplication theorem provided in Equation 5.1 to obtain $P(AB)$.

$$P(AB) = P = P(A)\ P(B\backslash A)$$

For case 1,

$$P = \left(\frac{5}{100}\right)\left(\frac{4}{99}\right)$$

$$= 0.002$$

For case 2,

$$P = \left(\frac{5}{100}\right)\left(\frac{5}{100}\right)$$

$$= 0.0025$$

COP 2: LIQUID AND SOLID LABORATORY SAMPLES

A laboratory contains 8 liquid samples and 11 solid samples. One sample is drawn at random without replacement; then, a second sample is also drawn at random.

1. What is the probability that both are liquid samples?
2. What is the probability that both are solid samples?

Solution

This is a conditional probability application. Let $P(L)$ be the probability of drawing a liquid sample, and $P(S)$ the probability of drawing a solid sample.

1. Here,

$$P = P(L)\ P(L\backslash L)$$

where $P(L\backslash L)$ is the probability that the second sample is liquid given that a liquid sample has been drawn.
It is noted that

$$P(L) = 8/19$$

$$P(L\backslash L) = 7/18$$

so that

$$P = \left(\frac{8}{19}\right)\left(\frac{7}{18}\right) = \frac{28}{171}$$
$$= 0.164$$

2. Here,

$$P = P(S)P(S\backslash S)$$
$$= \left(\frac{11}{19}\right)\left(\frac{10}{18}\right) = \frac{55}{171}$$
$$= 0.322$$

COP 3: Unusable Transistors

A supply of 5000 transistors contains usable transistors along with some transistors that may be repaired (R), some that can be discarded because of physical appearance (D), and some that could be used elsewhere (E). The potentially nonusable numbers are provided as follows:

R: 90
D: 25
E: 60
R and D: 6
R and E: 6
D and E: 4
R, D, and E: 2

If a nonusable transistor is obtained from the supply that is repairable, calculate the probability that the transistor can also be used elsewhere.

Solution

For this application,

$$P(RE) = P(R)\ P(E\backslash R)$$

Rearranging gives

$$P(E\backslash R) = \frac{P(RE)}{P(R)}$$

Based on the information provided,

$$P(R) = 90/5000$$

$$P(RE) = 6/5000$$

Therefore,

$$P(E \setminus R) = \frac{6/5000}{90/5000} = \frac{6}{90} = \frac{1}{15}$$

$$= 0.0667$$

COP 4: INCINERATOR "FINGERPRINTING"

Sampling of a liquid waste delivered to a hazardous waste incinerator is "fingerprinted" by first examining its color (C) and then its density (D) to determine if these properties are within previously set guidelines. Extensive data over a period of time provided the following results:

75% of the samples passed the color test
80% of those passing the color test also passed the density test
30% of those failing the color test passed the density test

For the next sample, calculate the probability of the following events:

1. It will pass the density test (the more important of the two tests).
2. It passes both the color and density tests.

Solution

The following events are defined:

C = pass color test
$\bar{C}$ = fail color test
D = pass density test

The events in the problem are now considered.

1. There are two tests for this case: passing both the density and color tests and passing the density test while failing the color test. Therefore,

$$P(D) = P(D \setminus C)P(C) + P(D \setminus \bar{C})P(\bar{C})$$

$$= (0.8)(0.75) + (0.3)(0.25)$$

$$= 0.6 + 0.075$$

$$= 0.675$$

2. The probability of passing the color test given that the density test has been passed is represented by $P(C|D)$:

$$P(C \setminus D) = \frac{P(CD)}{P(D)}$$

$$= \frac{P(D \setminus C)P(C)}{P(D)}$$

$$= \frac{(0.8)(0.75)}{(0.72)}$$

$$= 0.857$$

COP 5: Two Dice Calculations

Two dice are thrown. If the sum of the two dice is 8 (E), calculate the probability that one of the dice is a 2 (T).

Solution

Refer to Problem PDI 5 and Problem PDI 6 in Chapter 2. The sample space for E is given by

$$E = \{(2, 6), (6, 2), (3, 5), (5, 3), (4, 4)\}; \text{ 5 elements}$$

For T,

$$T = \{(1, 2), (2, 2), (2, 3), (2, 4), (2, 5), (2, 6), (2, 1), (3, 2), (4, 2),$$

$$(5,2), (6,2)\}; \text{ 11 elements}$$

For the two dice (TD), as before,

$$TD = \{36 \text{ elements}\}$$

For this throw,

$$P(\text{rolling a total of 8}) = P(E) = 5/36$$

$$P(\text{rolling a total of 8 and a 2}) = P(TE) = 2/36; (2, 6) \text{ and } (6, 2)$$

From Equation 5.1,

$$P(T \setminus E) = \frac{P(TE)}{P(E)}$$

$$= \frac{2/36}{5/36} = \frac{2}{5}$$

$$= 0.40$$

Note that this could have been deduced directly from the sample space of E.

COP 6: COMPANY SAFETY PROGRAM

At a chemical plant, the probability that a new employee who has attended the safety-training program will have a serious accident in his or her first year is 0.001. The corresponding probability for a new employee who has not attended the safety-training program is 0.02. If 80% of all new employees attend the safety-training program, what is the probability that a new employee will have a serious accident during the first year?

Solution

Let A be the event that a new employee attends the training program, and let B be the event that a new employee has a serious accident. In set notation,

$$B = AB + \bar{A}B$$

As AB and $\bar{A}B$ are mutually exclusive,

$$P(B) = P(AB) + P(\bar{A}B)$$

Application of the multiplication theorem yields

$$P(B) = P(A)P(B \backslash A) + P(\bar{A})P(B \backslash \bar{A})$$

$$= (0.80)(0.001) + (0.20)(0.02)$$

$$= 0.0048$$

COP 7: DEFECTIVE THERMOMETERS

From a supply of 300 thermometers, 20 of which are defective, 2 thermometers are drawn in succession. Find the probability that both thermometers are defective if the first thermometer is not replaced before the second is drawn.

Solution

Employ the following notation. Assign F for the first defective thermometer and S for the second defective thermometer. For this case,

$$P(S \backslash F) = \frac{P(SF)}{P(S)}$$

so that

$$P(SF) = P(F)\, P(F \backslash S)$$

Note that

$$P(S) = P(F) = 20/300$$

and

$$P(F \setminus S) = P(S \setminus F) = 19 / 299$$

$$P(SF) = \left(\frac{20}{300}\right)\left(\frac{19}{299}\right)$$

$$= 0.00424$$

COP 8: DEFECTIVE AUTOMOTIVE PARTS

Three automatic machines produce catalytic converters. Machine A produces 40% of the total, machine B, 25%, and machine C, 35%. On the average, 10% of the converters turned out by machine A do not conform to the specification; for machines B and C, the corresponding percentages are 5 and 1, respectively. If one part is selected at random from the combined output and is found not to conform to the specifications, what is the probability that it was produced by machine A?

Solution

If D represents a defective converter, the term to describe the desired answer is $P(A|D)$. Since

$$P(AD) = P(A \setminus D) \, P(D) = P(A) \, P(D \setminus A)$$

$$P(A/D) = \frac{P(A)P(D \setminus A)}{P(D)}$$

Based on the problem statement,

$$P(A) = 0.4$$

$$P(D \setminus A) = 0.1$$

To calculate $P(D)$, one notes

$$P(D) = P(DA) + P(DB) + P(DC)$$

$$P(D) = P(A) \, P(D \setminus A) + P(B) \, P(D \setminus B) + P(C) \, P(D \setminus C)$$

$$= (0.4)(0.1) + (0.25)(0.05) + (0.35)(0.01)$$

$$= 0.04 + 0.0125 + 0.0035$$

$$= 0.056$$

Therefore,

$$P(A \setminus D) = \frac{P(A)P(D \setminus A)}{P(D)}$$

$$= \frac{(0.4)(0.1)}{(0.056)}$$

$$= 0.714$$

This problem will be revisited in Chapter 10, Part II.

6 RAR. Random Variables and Random Numbers

INTRODUCTION

A random variable may be defined as a real-valued function defined over the sample space S of a random experiment. The domain of the function is S, and the real numbers associated with the various possible outcomes of the random experiment constitute the range of the function. If the range of the random variable consists of a finite number or countable infinitude of values, the random variable is classified as *discrete*. If the range consists of a noncountable infinitude of values, the random variable is classified as *continuous*. A set has a countable infinitude of values if they can be put into one-to-one correspondence with the positive integers. The positive even integers, for example, consist of a countable infinitude of numbers. The even integer $2n$ corresponds to the positive integer n for $n = 1, 2, 3, \ldots$. The real numbers in the interval $(0,1)$ constitute a noncountable infinitude of values.

Defining a random variable on a sample space S amounts to coding the outcomes in real numbers. Consider, for example, the random experiment involving the selection of an item at random from a manufactured lot. Associate $X = 0$ with the drawing of a nondefective item and $X = 1$ with the drawing of a defective item. Then X is a random variable with range $(0,1)$ and therefore discrete.

Let X denote the successive number of the throw on which the first failure of a switch occurs. Then X is a discrete random variable with range $\{1, 2, 3, \ldots, n, \ldots\}$. Note that the range of X consists of a countable infinitude of values and that X is therefore again discrete.

Suppose that X denotes the time to failure of a bus section in an electrostatic precipitator. Then X is a continuous random variable whose range consists of all the real numbers greater than zero.

The topic of random variables leads to the topic of random sampling, which in turn leads to the general subject of random numbers. In random sampling, each member of a population has an equal (the same) chance of being selected as a sample.

One technique available to ensure the selection of a random sample is through the use of random numbers. This process effectively reduces or eliminates the possibilities of encountering biased samples. There are several procedures available that one may employ to generate these random numbers. This is detailed in the following text.

As a simple application of random numbers, consider an environmental control device with time to failure T, measured in months, that has an exponential distribution with a probability distribution function (pdf) specified by

$$
\begin{aligned}
f(t) &= e^{-t}; & t > 0 \\
&= 0; & t \leq 0
\end{aligned}
$$

See the glossary (Appendix B) and Chapter 7 for an introduction to pdf. It is desired to estimate the average life of the device on the basis of a small number of simulated values of T.

By definition the cumulative distribution function (cdf) of T is obtained as follows:

$$F(t)=P(T \leq t)= \int_{0}^{t} f(t)dt$$

$$F(t)= \int_{0}^{t} e^{-t}dt = e^{-t}\Big|_{0}^{t} =-e^{-t} -(-e^{-0}) \qquad (6.1)$$

$$=-e^{-t}+1$$

Therefore,

$$F(t)=1-e^{-t}; \quad t>0$$
$$= 0; \qquad t \leq 0$$

If the first number generated by a random number generator in the 0 to 1 interval is 0.19, then the corresponding simulated value of T is obtained by solving the equation

$$0.19 = 1 - e^{-t}$$

to obtain 0.211. The process is repeated with another random number until the desired number of simulated values of T has been obtained. The average value of all the simulated values of T provides a random number estimate of the average life of the analytical device.

Monte Carlo simulation finds extensive application in applied statistics. This topic will be revisited in Chapter 20 of Part II of the text, particularly as it applies to a variety of distributions.

PROBLEMS AND SOLUTIONS

RAR 1: Random vs. Biased Sampling

Describe the difference between random sampling and biased sampling.

Solution

A *random* sample is a sample in which any one individual measurement in a population is as likely to be included as any other. Alternately, a *biased* sample is a sample in which certain individual measurements have a different chance to be included than others. The process of randomization, for example, through the use of random numbers, attempts to remove any bias in the sampling procedure.

RAR 2: Examples of Random Variables

Provide examples of random variables.

Solution

The following are a few examples of random variables:

1. The number of jacks in a pinochle hand
2. The number of aces in a bridge hand
3. The number of kings in a bridge hand

TABLE 6.1
The Die Probability Problem

X	1	2	3	> 3
P(X)	$\dfrac{1}{6}$	$\left(\dfrac{1}{6}\right)\left(\dfrac{5}{6}\right)$	$\left(\dfrac{1}{6}\right)\left(\dfrac{5}{6}\right)\left(\dfrac{5}{6}\right)$	$\dfrac{5^{X-1}}{6^X}$

---

4. The number of kings and queens in a bridge hand
5. The number of dots appearing on a die
6. The sum of dots (the number) appearing on a pair of dice
7. The time it takes for a pump to fail

RAR 3: RANDOM VARIABLE OF ONE DIE

Express the probability that one dot will appear on a thrown die in terms of the number of times the die is tossed.

Solution

Let X be a random variable denoting the number of times a die is thrown until a one appears. The sample space of X is (once again)

$$X = \{1, 2, 3, \ldots\}$$

The probability of X is provided in Table 6.1.
This result may also be shown in equation form.

$$P(X) = \frac{5^{X-1}}{6^X}; \quad X = 1, 2, 3, \ldots$$

RAR 4: RANDOM NUMBER VERIFICATION

Consider the following numbers:

$$8, 14, 23, 34, 42, 49, 57, 66$$

Comment on whether these are random numbers, a result of a set progression, or the results of an analytical expression or whether there is some other explanation for this sequence.

Solution

Care should be exercised in interpreting numbers. The numbers given in the preceding statement represent local train stops on the once designated Brooklyn-Manhattan Transit (BMT) Corporation subway in New York City. One of the authors of this text rode this subway for 4 years to the prestigious Cooper Union School of Engineering 5 days a week over 50 years ago.

RAR 5: FOOTBALL POOL

One of the major gambling options during the professional football championship game (Super Bowl) is to "buy a box" in a uniquely arranged square, usually referred to as the *pool*. An example of a pool is shown in Figure 6.1.

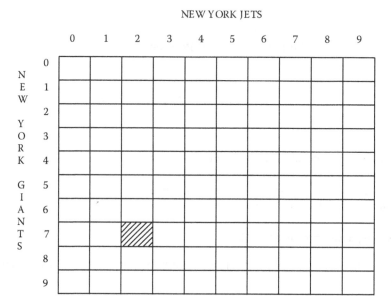

FIGURE 6.1 Nonrandom pool.

As can be seen, there are 100 boxes. If each box costs $1,000, the total cash pool is $100,000. The individual, who selects the box with the last digit of the final score for each team takes home the bacon, i.e., wins the $100,000. If the final score is Jets 22/Giants 7, the owner of the shaded box is the winner. Scores such as Jets 12/Giants 27 or Jets 22/Giants 37 would also be winners.

Explain why employing this format does not provide each person buying a box an equal chance of winning.

Solution

Knowledgeable football fans would immediately realize that the best numbers to select are 0 and 7, whereas the worst are 2, 5, and 8. Therefore, the arrangement of the box as in Figure 6.1 does not provide each bettor with an equal chance to win. This bias can be removed by assigning the numbers to each team in a random manner after individuals have paid and selected a box. The procedure most often used is to write numbers from 0 to 9 (each) on a piece of paper. The ten pieces of paper are then randomly drawn from a container and sequentially placed along the side of the square — first horizontally and then vertically. The result might look like Figure 6.2.

Had the same box been selected (see Figure 6.1), the bettor's last digit would be Giants 2/Jets 0. A final score of Giants 42/Jets 30 would be a winner as would Giants 12/Jets 20.

RAR 6: COMPUTER TESTING

Approximately 2000 computers are produced daily at a plant and are numbered sequentially, starting with 1, as they come off the assembly line. Once a month on the first Monday, eight computers are randomly selected and thoroughly tested. Last month, 1928 computers were produced. Outline how the computer production line could be checked to negate any variation in quality with time of day.

Solution

This problem can be solved using random numbers. Random number generators are available that can produce near infinite numbers containing 4 digits, e.g., 1039; tables are also available that

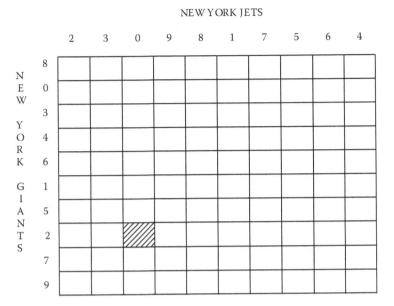

FIGURE 6.2 Random pool.

TABLE 6.2
Random Four-Digit Numbers

9311	3966	1880	4227	1728	6714	9077	7123
4286	4097	1211	0977	9441	3217	1666	7396
0085	6322	6322	9004	9181	4890	0325	3944
3215	9908	7987	0426	0555	7211	8890	1828

contain multidigit random numbers that are free from bias. Examples of 4-digit random numbers are provided in Table 6.2.

Based on Table 6.2, only the numbers that are boxed would be eligible for consideration because they are located in the 1 to 1928 or 0001 to 1928 range. Therefore, the eight computers selected for checking would be numbered 85, 1880, 1211, 977, 426, 1728, 555, and 1666. Note that numbers greater than 1928 are automatically rejected, and only the first 8 eligible 4-digit numbers are selected. Also, note that repeaters should be discarded.

RAR 7: MEMBERSHIP DIRECTORY

The Air & Waste Management Association (AWMA) membership directory contains 611 pages. Each page has 90 lines (2 columns of 45 lines), and each individual requires between 4 to 6 lines; however, each individual's name is contained on one line. It has been proposed to send questionnaires to 50 members. Outline a procedure to randomly select the 50 members for the poll.

Solution

This solution requires generating two sets of random numbers — one with three digits and the other with two digits. The procedure is as follows. Obtain the first three-digit random number. If the number is above 611, discard it, and obtain another number. If the number is below 612, proceed to that page, e.g., 497. Now select a two-digit number. If that number is above 90, discard it, and obtain another 2-digit number. If the number is below 91, proceed to that line on page 497. If a name does not appear on that line, select another 2-digit number until the number is below 91, and the line corresponding to the number contains a name, e.g., 17. This is the first member randomly selected. The process is repeated until 50 names are obtained.

A simpler procedure might be available. If the 9036 members are listed alphabetically and numbered sequentially, one could generate 4-digit numbers (see Problem RAR 6) and select the first 50 that are below 9037, discarding repeaters.

RAR 8: CABLE WIRE MEASUREMENT LOCATION

The diameter of an 8-mile-long cable wire is to be measured. A total of 25 measurements are to be made at various locations along the cable. Describe how to select these locations using random numbers if the sampling is 1/2 ft long.

Solution

The length (L) of cable in feet is $L = (5280)(8) = 42,240$ ft. If the cable's full length is divided into 1/2 ft increments, a total of 84,480 increments would result. The locations for diameter sampling would be obtained by generating 5-digit numbers between 0 and 99,999 and selecting the first 25 that are in the 0 to 84,480 range.

RAR 9: CONTAMINATED SOIL

The contaminated soil at a Superfund site is to be sampled at a number of locations. If the surface area of the site is rectangular in shape, suggest the location where the measurement should be made in order that the measurements are representative of the site.

Solution

A suggested procedure is to have the two-dimensional surface laid out in a large number of small equal areas, the center of each being the point where a measurement is to be taken. For rectangular surfaces, the area is divided into equal areas of the same shape with the sampling point located at the center of each equal area, as shown in Figure 6.3.

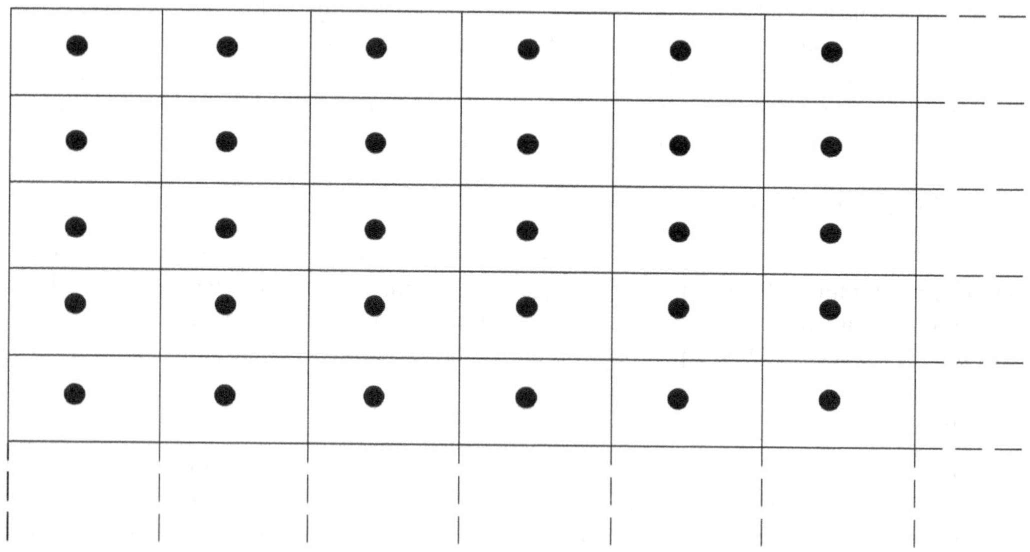

FIGURE 6.3 Sampling locations.

TABLE 6.3
Time to Failure Results

Random Number	Simulated Time to Failure (years)
0.93	2.66
0.06	0.06
0.53	0.76
0.56	0.82
0.41	0.53
0.38	0.48
0.78	1.52
0.54	0.78
0.49	0.67
0.89	2.21
0.77	1.47
0.85	1.90
0.17	0.19
0.34	0.42
0.56	0.82

Depending on the number of rectangles in each of the two perpendicular directions, two random numbers are generated that would locate the sampling site on the grid. This process is repeated until an appropriate number of sampling sites is specified (located).

RAR 10: Time to Failure: Analytical Device

Refer to the time to failure of the device discussed in the introductory section of this chapter. Perform 15 calculations to estimate the average life of the device.

Solution

For each calculation, a random number (RN) between 0 and 1, to 2 significant digits, is used to calculate the time to failure (t) in the equation

$$RN = 1 - e^{-t}$$

The results are provided in Table 6.3.

The average of the 15 runs is 1.02 years and represents the estimate of the average life of the analytical device. A more accurate estimate of the true value of the device can be obtained by increasing the number of simulated values on which the estimate is based. This problem will be revisited in Chapter 16 of Part II.

RAR 11: ELECTROSTATIC PRECIPITATOR: BUS SECTION FAILURES

An electrostatic precipitator (ESP) is a particle control device that uses electrical forces to move the particles out of a flowing gas stream and onto collector plates. The particles are given an electric charge by forcing them to pass through a corona, a region in which gaseous ions flow. The electrical field that forces the charged particles to the walls results from electrodes maintained at high voltage in the center of the flow lane. Once the particles are collected on the plates, they must be removed from the plates without re-entraining them into the gas stream. This is usually accomplished by

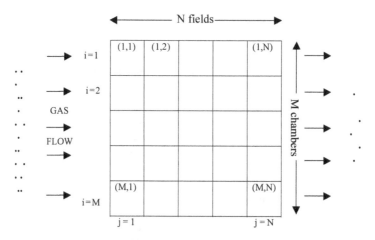

FIGURE 6.4 An M × N precipitator (top view).

their being dislodged from the plates, the collected layer of particles being allowed to slide down under the influence of gravity into a hopper, from which they are removed. This is usually the environmental control unit of choice for coal-fired boilers in the utility industry.

One of the authors has employed Monte Carlo methods to estimate out-of-compliance probabilities for electrostatic precipitators on the basis of observed bus section failures. The following definitions apply (see also Figure 6.4).

Chamber	One of many passages for gas flow.
Field	One of several high voltage sections for the removal of particulates; these fields are arranged in series (i.e., the gas passes from the first field into the second, etc.).
Bus section	A region of the precipitator that is independently energized; a given bus section can be identified by a specific chamber and field.

The beneficial effect of sectionalization of electrostatic precipitators is well known. Particle collection efficiency can be expected to increase as the number of independently energized sections increases.

During operation, a precipitator is "out of compliance" when its overall collection efficiency falls below a designated minimum because of bus section failures. The overall collection efficiency is calculated from the individual bus section efficiency; a procedure that is complicated because the location of the bus section affects the efficiency calculation. This procedure is beyond the scope of this text. When several bus sections fail, the effect of the failures depends on where they are located.

Outline how to determine whether a precipitator is out of compliance after a given number of bus sections have failed.

Solution

To determine directly whether a precipitator is out of compliance after a given number of bus sections have failed, it would be necessary to test all possible arrangements of the failure locations. The out-of-compliance probability is given by the percentage of arrangements that result in overall collection efficiencies less than the prescribed minimum standard. The number of arrangements to be tested often makes the direct approach impractical. For example, the author in question was once requested to investigate a precipitator unit consisting of 64 bus sections; if 4 of these were to fail, there would be 15,049,024 possible failure arrangements.

Instead of the direct approach that would have required study of more than 15×10^6 potential failure arrangements, the author used a Monte Carlo technique, testing only a random sample of possible failure arrangements. The arrangements to be chosen were selected by use of random numbers, as described in the following text.

A set of random numbers is generated equal in quantity to the number of bus section failures assumed. Each of the random numbers is used to identify a bus section, which, during the calculation of overall collection efficiency, is assumed to be out of commission. A total of 5000 failure location arrangements were sampled for all the out-of-compliance probabilities calculated. The random numbers used were generated by the power residue method.

RAR 12: Two-Dimensional Heat Conduction

Consider the two-dimensional (y,z) heat transfer problem of conduction in a solid presented in Figure 6.5.

The top surface of the square rod is in a steam bath at 100°C, whereas the bottom and two vertical surfaces are maintained at 0°C (an ice bath). Provide three methods of solving for the temperature profile. (One of the authors has employed this problem as a project in a heat transfer class; however, numerical solutions were also required). Detail the random number approach.

Solution

The system given in the preceding text is described by a partial differential equation.

$$\frac{\partial^2 T}{\partial y^2} + \frac{\partial^2 T}{\partial z^2} = 0$$

Two boundary conditions in y and two in z are required for the solution, and these are specified. The three methods of solution are:

1. Analytical using separation of variables
2. Numerical using finite difference
3. Monte Carlo simulation employing random numbers

For method 3, the solution requires a large number of simple calculations. Consider the point A in Figure 6.5. The procedure entails generating 2-digit numbers (00 to 99). If the 2-digit number is in the 0 to 24 range, move horizontally from A to the left to the next intersection; for 24 to 49,

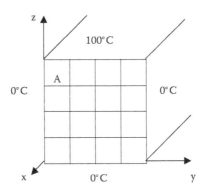

FIGURE 6.5 Heat conduction in a solid.

move to the right; for 50 to 74, move vertically upward; and for 75 to 99, move downward. With each two-digit number, determine if the intersection is at an outer surface. When an outer surface is reached, the temperature is recorded, and another trial is initiated. This process is repeated, say, 5000 times; the sum of the temperatures recorded for the 5000 trials is then divided by 5000. The answer ultimately results in a temperature of approximately 93°C for point A. The procedure is repeated for the remaining points to generate a temperature profile for the solid. This problem is revisited in more detail in Chapter 20, Problem MON 6.

7 PDF. Introduction to Probability Distributions

INTRODUCTION

The probability distribution of a random variable concerns the distribution of probability over the range of the random variable. The distribution of probability, i.e., the values of the discrete random variables together with their associated probabilities, is specified by the *probability distribution function* (pdf). This chapter is devoted to providing general properties of the pdf in the case of discrete and continuous random variables, as well as an introduction to *cumulative distribution function* (cdf). Special pdfs finding extensive application in hazard and risk analysis are considered in Section II.

The pdf of a discrete random variable X is specified by $f(x)$, where $f(x)$ has the following essential properties:

1. $f(x) = P(X = x) =$ probability assigned to the outcome corresponding (7.1)
    to the number x in the range of X; i.e., X is a specifically designated value of x
2. $f(x) \geq 0$ (7.2)

3. $\sum_{x} f(x) = 1$ (7.3)

Property 1 indicates that the pdf of a discrete random variable generates probability by substitution. Property 2 and Property 3 restrict the values of $f(x)$ to nonnegative real numbers and numbers whose sum is 1, respectively.

Consider, for example, a box of 100 transistors containing 5 defectives. Suppose that a transistor selected at random is to be classified as defective or nondefective. Then X is a discrete random variable with pdf specified by

$$f(x) = 0.05; \quad x = 1$$

$$= 0.95; \quad x = 0$$

For another example of the pdf of a discrete random variable, let X denote the number of the throw on which the first failure of an electrical switch occurs. Suppose the probability that a switch fails on any throw is 0.001 and that successive throws are independent with respect to failure. If the switch fails for the first time on throw x, it must have been successful on each of the preceding $x - 1$ trials. In effect, the switch survives up to $x - 1$ trials and fails at trial x. Therefore, the pdf of X is given by

$$f(x) = (0.999)^{x-1}(0.001); \quad x = 1, 2, 3, \ldots, n, \ldots$$

Note that the range of X consists of a countable infinitude of values. (See also Problem PDF 12 in Chapter 7.)

The pdf of a continuous random variable X has the following properties:

1. $\displaystyle\int_a^b f(x)\,dx = P(a < X < b)$ (7.4)

2. $f(x) \geq 0$ (7.5)

3. $\displaystyle\int_{-\infty}^{\infty} f(x)\,dx = 1$ (7.6)

Equation 7.4 indicates that the pdf of a continuous random variable generates probability by integration of the pdf over the interval whose probability is required. When this interval contracts to a single value, the integral over the interval becomes zero. Therefore, the probability associated with any particular value of a continuous random variable is zero. Consequently, if X is continuous

$$P(a \leq X \leq b) = P(a < X \leq b)$$

$$= P(a < X < b) \tag{7.7}$$

$$= P(a \leq X < b)$$

Equation 7.5 restricts the values of $f(x)$ to nonnegative numbers. Equation 7.6 follows from the fact that

$$P(-\infty < X < \infty) = 1 \tag{7.8}$$

As an example of the pdf of a continuous random variable, consider the pdf of the time X in hours between successive failures of an aircraft air conditioning system. Suppose the pdf of X is specified by

$$f(x) = 0.01\, e^{-0.01x}; \quad x > 0$$

$$= 0; \quad \text{elsewhere}$$

A plot of $f(X)$ vs. X for positive values of X is provided in Figure 7.1. Inspection of the graph indicates that intervals in the lower part of the range of X are assigned greater probabilities than intervals of the same length in the upper part of the range of X because the areas over the former are greater than the areas over the latter.

The expression $P(a < X < b)$ can be interpreted geometrically as the area under the pdf curve over the interval (a,b). Integration of the pdf over the interval yields the probability assigned to the interval. For example, the probability that the time in hours between successive failures of the aforementioned aircraft air conditioning system is greater than 6 but less than 10 is

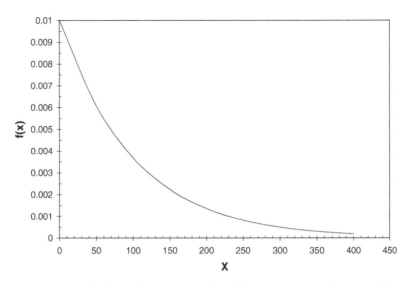

FIGURE 7.1 The pdf of time in hours between successive failures of an aircraft air conditioning system.

$$P(6 < X < 10) = \int_6^{10} 0.01 e^{-0.01x} dx = 0.01 \int_6^{10} e^{-0.01x} dx$$

$$= 0.01 \left[-\left(\frac{1}{0.01} \right) e^{-0.01x} \right]_6^{10}$$

$$= -e^{-0.01x} \Big|_6^{10}$$

$$= -e^{-(0.01)10} - (-e^{-(0.01)6})$$

$$= -e^{-0.1} + e^{-0.06}$$

$$= -0.9048 + 0.9418$$

$$= 0.037$$

Another function used to describe the probability distribution of a random variable X is the cdf. If $f(x)$ specifies the pdf of a random variable X, then $F(x)$ is used to specify the cdf. For both discrete and continuous random variables, the cdf of X is defined by

$$F(x) = P(X \geq x); \quad -\infty < x < \infty \tag{7.7}$$

Note that the cdf is defined for all real numbers, not just the values assumed by the random variable.

To illustrate the derivation of the cdf from the pdf, consider the case of a random variable X whose pdf is specified by

$$f(x) = 0.2; \quad x = 2$$

$$= 0.3; \quad x = 5$$

$$= 0.5; \quad x = 7$$

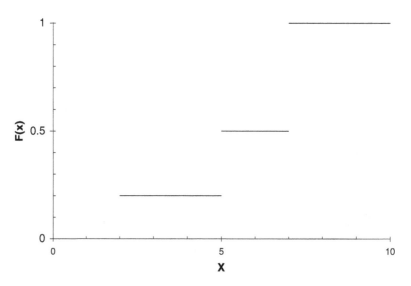

FIGURE 7.2 Graph of the cdf of a discrete random variable X.

Applying the definition of the cdf in Equation 7.7, one obtains for the cdf of X (see also Figure 7.2).

$$F(x) = 0; \quad x < 2$$

$$= 0.2; \quad 2 \le x < 5$$

$$= 0.5; \quad 5 \le x < 7$$

$$= 7; \quad x \ge 7$$

It is helpful to think of $F(x)$ as an accumulator of probability as x increases through all real numbers. In the case of a discrete random variable, the cdf is a step function increasing by finite jumps at the values of x in the range of X. In the aforementioned example, these jumps occur at the values 2, 5, and 7. The magnitude of each jump is equal to the probability assigned to the value at which the jump occurs. This is depicted in Figure 7.2.

In the case of a continuous random variable, the cdf is a continuous function. Suppose, for example, that X is a continuous random variable with pdf specified by

$$f(x) = 2x; \quad 0 \le x < 1$$

$$= 0; \quad \text{elsewhere}$$

Applying the definition once again, one obtains

$$F(x) = 0; \qquad\qquad x < 0$$

$$= \int_0^x 2x\,dx = x^2; \quad 0 \le x < 1$$

$$= 1; \qquad\qquad x \ge 1$$

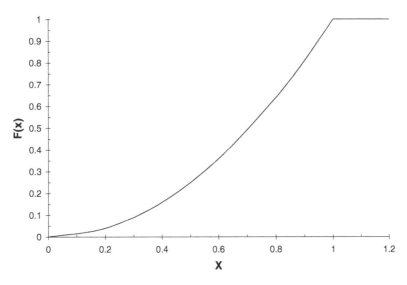

FIGURE 7.3 Graph of the cdf of a continuous random variable X.

Figure 7.3 displays the graph of this cdf, which is simply a plot of $f(X)$ vs. X. Differentiating cdf and setting pdf equal to zero where the derivative of cdf does not exist can provide the pdf of a continuous random variable. For example, differentiating the cdf of X^2 yields the specified pdf of $2X$. In this case, the derivative of cdf does not exist for $x = 1$.

The following properties of the cdf of a random variable X can be deduced directly from the definition of $F(x)$.

1. $F(b) - F(a) = P(a < X \le b)$ (7.8)
2. $F(+\infty) = 1$ (7.9)
3. $F(-\infty) = 0$ (7.10)
4. $F(x)$ is a nondecreasing function of X   (7.11)

These properties apply to the cases of both discrete and continuous random variables.

PROBLEMS AND SOLUTIONS

PDF 1: Discrete vs. Continuous Random Variables

Provide an example of a *discrete* random variable and a *continuous* random variable.

Solution

A random variable may be *discrete*; i.e., it can take on only a finite or countable number of distinct values. Alternately, the random variable may be *continuous*, with values occurring anywhere in an interval. The outcome of the throw of a die is a discrete random variable. The exact height of an individual selected from a particular group is a continuous variable.

PDF 2: Differences Between pdf and cdf

Describe how the pdf and cdf for a random discrete variable differ from that of a random continuous variable.

TABLE 7.1
Sample Probability Distribution

X	x_1	x_2	...	x_i
$P(X)$	$P(x_1)$	$P(x_2)$	...	$P(x_i)$

TABLE 7.2
Two-Dice Probability Distribution

X	2	3	4	5	6	7	8	9	10	11	12
$P(X)$	1/36	2/36	3/36	4/36	5/36	6/36	5/36	4/36	3/36	2/36	1/36

Solution

As discussed earlier in the introductory section of this chapter, pdf and cdf are analogous except that the integrals are replaced by summations over points of the sample space. In this case, the pdf directly provides the probability attached to each discrete value of the random variable.

PDF 3: DISCRETE VARIABLE AND PDF GENERALIZATION

Provide a general definition in notation form of the pdf of a discrete random variable. Also include an example involving a dice game.

Solution

Suppose X can take on only the values $x_1, x_2, ..., x_i$ with corresponding probabilities $P(x_1)$, $P(x_2)$, ..., $P(x_i)$. For the random variable X, one can write

$$P(x_1) + P(x_2) + ... + P(x_i) = 1$$

These values of x and the corresponding probabilities can be arranged in table form (see Table 7.1).

As noted earlier, the values of the discrete random variable X together with their associated probabilities are called the *probability distribution function* (pdf) of X.

As an example, let X denote the sum of the points obtained on throwing two dice. The probability distribution is given in Table 7.2.

Thus, the probability of rolling a 7 (as shown earlier) is 6/36 or 1/6.

PDF 4: SUPPLY SHIP DELIVERY

Let X denote the number of supply ships reaching a port on a given day. The pdf for X is given by:

$$f(x) = \frac{x}{21}; \quad x = 1, 2, 3, 4, 5, 6$$

1. Verify that $f(x)$ is a valid pdf.
2. Calculate the probability that at least three ships but less than six ships will arrive on a given day.

Solution

1. For $f(x)$ to be valid, the following two conditions must be satisfied:

$$0 \leq f(x) \leq 1; \quad x \text{ between 1 and 6}$$

and

$$\sum_{1}^{6} f(x) = 1$$

Substituting into the second condition yields

$$\sum_{1}^{6} f(x) = \sum_{1}^{6} \left(\frac{x}{21} \right)$$

$$= \frac{1}{21} + \frac{2}{21} + \frac{3}{21} + \frac{4}{21} + \frac{5}{21} + \frac{6}{21}$$

$$= \frac{21}{21} = 1.0$$

Thus, both conditions are satisfied and $f(x)$ is a valid pdf.

2. The probability that at least 3 ships but no more than 6 ships will arrive on a given day can be determined by

$$P(3 \leq X < 6) = f(x = 3) + f(x = 4) + f(x = 5)$$

$$= \frac{3}{21} + \frac{4}{21} + \frac{5}{21}$$

$$= \frac{12}{21} = 0.429$$

PDF 5: EMPLOYEE RETENTION

Let the discrete random variable X represent the number of years before an employee at an environmental consulting firm leaves the company. The pdf of X is given by

$$f(x) = \frac{k}{x}; \quad x = 1, 2, 3, 4$$

1. Find the constant k.
2. Plot the cdf, $F(x)$, as a function of X.

Solution

1. Assuming a valid pdf,

$$\sum_{1}^{4} f(x) = \sum_{1}^{4}\left(\frac{k}{x}\right) = 1.0; \quad x = 1, 2, 3, 4$$

Substituting gives

$$k + \frac{k}{2} + \frac{k}{3} + \frac{k}{4} = 1.0$$

Solving for k,

$$k = \frac{12}{25} = 0.48$$

2. To generate a cdf plot, one notes,

$$F(x) = P(X \leq x) = \sum_{1}^{x} f(x); \quad x = 1, 2, 3, \ldots, x$$

The following values are calculated,

$$F(1) = f(1) = \frac{12}{25}$$

$$F(2) = f(1) + f(2) = \frac{12}{25} + \frac{6}{25} = \frac{18}{25}$$

$$F(3) = f(1) + f(2) + f(3) = \frac{12}{25} + \frac{6}{25} + \frac{4}{25} = \frac{22}{25}$$

$$F(4) = f(1) + f(2) + f(3) + f(4) = \frac{12}{25} + \frac{6}{25} + \frac{4}{25} + \frac{3}{25} = \frac{25}{25} = 1.0$$

The cdf plot is provided in Figure 7.4.

PDF 6: ACCIDENT CLAIMS

A random variable X, denoting the years of college education of an employee selected at random from the personnel of an insurance company dealing with health-related accident claims has the pdf

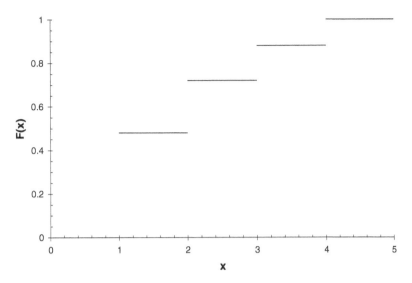

FIGURE 7.4 Employee retention cdf plot.

$$f(x) = 0.6; \quad x = 2$$

$$= 0.4; \quad x = 4$$

Find the cdf of X.

Solution

Because

$$F(x) = P(X \leq x)$$

the cdf of X is given by

$$F(x) = 0; \quad x < 2$$

$$= 0.6; \quad 2 \leq x < 4$$

$$= 1.0; \quad x \geq 4$$

PDF 7: ANNUAL FLOODS

Let X denote the annual number of floods in a certain region. The pdf of X is specified as

$$f(x) = 0.25; \quad x = 0$$

$$f(x) = 0.35; \quad x = 1$$

$$f(x) = 0.24; \quad x = 2$$

$$f(x) = 0.11; \quad x = 3$$

$$f(x) = 0.04; \quad x = 4$$

$$f(x) = 0.01; \quad x = 5$$

1. What is the probability of two or more floods in any year?
2. What is the probability of four or more floods in a year, given that two floods have already occurred?
3. What is the probability of at least one more flood in a year, given that one has already occurred?

Solution

1. For two or more floods in any year, one may write

$$P(x \geq 2) = \sum_{2}^{\infty} f(x) = \sum_{2}^{5} f(x)$$

$$= f(x = 2) + f(3) + f(4) + f(5)$$

$$= 0.24 + 0.11 + 0.04 + 0.01$$

$$= 0.40$$

2. This involves a conditional probability calculation. The describing equation and solution is

$$P(X \geq 4 \backslash X \geq 2) = P(X \geq 4 \text{ and } X \geq 2)/P(X \geq 2)$$

$$= P(X \geq 4)/P(X \geq 2)$$

$$= 0.05/0.40$$

$$= 0.125$$

3. This also involves a conditional probability calculation. For this case,

$$P(X \geq 2 \backslash X \geq 1) = P(X \geq 2 \text{ and } X \geq 1)/P(X \geq 1)$$

$$= P(X \geq 2)/P(X \geq 1)$$

$$= 0.40/(0.40 + 0.35)$$

$$= 0.40/0.75$$

$$= 0.53$$

PDF 8: CONTINUOUS RANDOM VARIABLE

A continuous random variable X has a pdf of $X/2$ for $0 \leq x \leq 2$. What is the cdf?

Solution

By definition, for continuous variables,

$$F(x) = \int f(x)\, dx = \int \frac{x}{2}\, dx$$

$$= \frac{x^2}{4}; \quad 0 \le x < 2$$

Applying the definition of cdf leads to

$$F(x) = 0; \quad x < 0$$

$$F(x) = \frac{x^2}{4}; \quad 0 \le x < 2$$

$$F(x) = 1; \quad x \ge 2$$

PDF 9: ORIGINAL STUDENT PDF

An environmental student recently submitted the following original problem. Let X be a continuous random variable with a pdf given by

$$f(x) = x^2 + kx + 3; \quad 0 \le x \le 4$$

Determine the constant k such that $f(x)$ is a valid pdf.

Solution

The constant k must satisfy two conditions:

1. $\int_0^4 f(x)\, dx = 1$
2. $f(x) \ge 0; \quad 0 \le x \le 4$

For Condition 1,

$$\int_0^4 (x^2 + kx + 3)\, dx = 1$$

$$= \frac{x^3}{3} + \frac{kx^2}{2} + 3x = 1$$

$$= \frac{64}{3} + 8k + 12 = 1$$

Solving for k yields

$$k = -\frac{97}{24}$$

For Condition 2, determine $f(x)$ vs. x and ascertain whether $f(x)$ is positive over the range $0 \leq x \leq 4$. For $x = 3$;

$$f(x) = 3^2 - \left(\frac{97}{24}\right)(3) + 3$$

$$= \text{negative value}$$

Because $f(x)$ is not positive for values of x between 0 and 4, the pdf provided by the student is not valid with the calculated value of k. A flunking grade may be in order.

PDF 10: RICHTER SCALE EARTHQUAKE MEASUREMENTS

The difference between the magnitude of a large earthquake, as measured on the Richter scale, and the threshold value of 3.25 is a random variable X having the pdf

$$f(x) = 1.7\,e^{-1.7x}; \quad x > 0$$

$$= 0; \qquad \text{elsewhere}$$

Calculate $P(2 < X < 6)$.

Solution

Applying the definition of cdf leads to

$$P(2 < X < 6) = \int_2^6 f(x)\,dx$$

Substituting and integrating yields

$$P(2 < X < 6) = \int_2^6 1.7e^{-1.7x}\,dx$$

$$= e^{-3.4} - e^{-10.2}$$

$$= 0.0334 - 0.0$$

$$= 0.0334$$

PDF 11: Battery Life

A random variable X denoting the useful life of a battery in years has the pdf

$$f(x) = \left(\frac{3}{8}\right)x^2; \quad 0 < x < 2$$

$$= 0; \qquad \text{elsewhere}$$

Find the cdf of X.

Solution

As before, the cdf of X is given by

$$f(x) = P(X \leq x)$$

$$= \int_{-\infty}^{x} f(x)\,dx$$

Therefore,

$$F(x) = 0; \qquad x \leq 0$$

$$F(x) = \int [(3/8)x^2]dx$$

$$= \frac{x^3}{8}; \qquad 0 < x < 2$$

$$F(x) = 1; \qquad x \geq 2$$

PDF 12: Light Switch Failure

The probability that a light switch fails is 0.0001. Let X denote the trial number on which the first failure occurs. Find the probability that X exceeds 1000.

Solution

The probability of no failures on the first $x - 1$ trials and failure on trial x is $(0.9999)^{x-1}(0.0001)$. Therefore, the pdf of X is

$$f(x) = (0.9999)^{x-1}(0.0001)$$

which is a geometric series with the first term equal to

$$f(x) = (0.9999)^{1000}(0.0001)$$

and common ratio equal to 0.9999.

This reduces to

$$P(X > 1000) = (0.9999)^{1000}(0.0001)/(1 - 0.9999)$$

$$= 0.9048$$

8 EXV. Expected Values

INTRODUCTION

By definition, the expected value of a random variable is the average value of the random variable. The expected value of a random variable X is denoted by $E(X)$. The expected value of a random variable can be interpreted as the long-run average of observations on the random variable. The procedure for calculating the expected value of a random variable depends on whether the random variable is discrete or continuous.

If X is a discrete random variable with probability distribution function (pdf) specified by $f(x)$, then

$$E(X) = \sum_x xf(x)$$

or

$$E[g(X)] = \sum_x g(x)f(x) \tag{8.1}$$

If X is a continuous random variable with pdf specified by $f(x)$, then

$$E(X) = \int_{-\infty}^{\infty} xf(x)\,dx$$

or

$$E[g(X)] = \int_{-\infty}^{\infty} g(x)f(x)\,dx \tag{8.2}$$

Additional rules concerning expected values are as follows:

1. The expected value of a constant is the constant itself; i.e.,

$$E(c) = c; \quad c \text{ is any constant} \tag{8.3}$$

This can be verified by noting that

$$E(c) = \int_{-\infty}^{\infty} cf(x)\,dx = c\int_{-\infty}^{\infty} f(x)\,dx = c$$

Because (by definition)

$$\int_{-\infty}^{\infty} f(x)\,dx = 1.0$$

2. The expected value of a constant times a random variable is the constant times the expected value of the random variable; i.e.,

$$E(cX) = cE(X) \tag{8.4}$$

This can also be verified by noting that

$$E(cX) = \int_{-\infty}^{\infty} cxf(x)\,dx = c\int_{-\infty}^{\infty} xf(x)\,dx = cE(X)$$

3. The expected value of two terms is the sum of the expected value of each; i.e.,

$$E[f(x) + g(x)] = E[f(x)] + E[g(x)] \tag{8.5}$$

Suppose, for example, that the pdf of a discrete random variable X is specified by

$$f(x) = 0.3; \quad x = 10$$

$$= 0.2; \quad x = 20$$

$$= 0.5; \quad x = 30$$

By Equation 8.1, the expected value of X is given by

$$E(X) = 10(0.3) + 20(0.2) + 30(0.5) = 22$$

Suppose, for example, that the pdf of a continuous random variable x is specified by

$$f(x) = e^{-x}; \quad x > 0$$

$$= 0; \quad \text{elsewhere}$$

Application of Equation 8.2 yields the following expected value for X.

$$E(X) = \int_{0}^{\infty} xe^{-x}\,dx = 1$$

The expected value of a random variable X is also called *the mean of X* and is often designated by μ. The expected value of $(X - \mu)^2$ is called the *variance* of X. The positive square root of the variance is called the *standard deviation*. The terms σ^2 and σ (sigma squared and sigma) represent the variance and standard deviation, respectively. Variance is a measure of the spread or dispersion of the values of the random variable about its mean value. The standard deviation is also a measure of spread or dispersion. The standard deviation is expressed in the same units as X, whereas the variance is expressed in the square of these units.

The variance of X can be calculated directly from the following definition:

$$\sigma^2 = E(X - \mu)^2 \tag{8.6}$$

However, it is usually calculated more easily by the equivalent formula

$$\sigma^2 = E(X^2) - \mu^2 = E(X^2) - [E(X)]^2 \tag{8.7}$$

To illustrate the computation of variance and its interpretation in the case of discrete random variables, consider a random variable X having pdf specified by

$$f(x) = 0.5; \quad x = 1$$

$$= 0.5; \quad x = -1$$

and a random variable Y having pdf specified by

$$g(y) = 0.5; \quad y = 10$$

$$= 0.5; \quad y = -10$$

It is easily verified that both X and Y have the same expected value of zero, i.e., $E(X) = E(Y) = 0$. From Equation 8.1, the expected value of X^2 is given by

$$E(X^2) = \sum_x x^2 f(x)$$

$$= (1)(0.5) + (1)^2(0.5) = 1$$

From Equation 8.7, the variance of X is 1. The expected value of Y^2 is given by

$$E(Y^2) = \sum_y y^2 g(y)$$

$$= (10)^2(0.5) + (10)^2(0.5) = 100$$

Therefore, the variance of Y is 100. Furthermore, the standard deviation of X is 1 and the standard deviation of Y is 10. The larger value for the standard deviation in the case of Y reflects the greater dispersion of Y values about their mean.

To illustrate the computation of variance and its interpretation in the case of continuous random variables, consider a random variable X having a pdf specified by

$$f(x) = \frac{1}{2}; \quad 0 < x < 2$$

$$= 0; \quad \text{elsewhere}$$

and a random variable Y having a pdf specified by

$$g(y) = \frac{1}{4}; \quad -1 < y < 3$$

$$= 0; \quad \text{elsewhere}$$

The mean value of X is again given by Equation (8.2):

$$E(X) = \int_0^2 x\left(\frac{1}{2}\right) dx$$

$$= 1$$

The expected value of X^2 is given by

$$E(X^2) = \int_0^2 x^2\left(\frac{1}{2}\right) dx$$

$$= \frac{4}{3}$$

From Equation 8.7, the variance of X is

$$\sigma^2 = E(X^2) - \mu^2$$

$$= \frac{4}{3} - 1 = \frac{1}{3}$$

The mean value of Y is given by

$$E(Y) = \int_{-1}^3 y\left(\frac{1}{4}\right) dy$$

$$= 1$$

The expected value of Y^2 is given by

$$E(Y^2) = \int_{-1}^3 y^2\left(\frac{1}{4}\right) dy$$

$$= \frac{7}{3}$$

Therefore, the variance of Y is

$$\sigma^2 = E(Y^2) - \mu^2$$

$$= \frac{7}{3} - 1 = \frac{4}{3}$$

Here, X and Y have the same expected values, but Y has the larger variance, once again reflecting the greater dispersion of Y values about their mean.

PROBLEMS AND SOLUTIONS

EXV 1: EXPECTED VALUE DEFINITION

Describe the term expected value (or expectation) in terms of probability. Also, include an example.

Solution

Let X represent a discrete random variable and P its associated probabilities. The expected value of X, $E(X)$, is given by

$$E(X) = P_1 X_1 + P_2 X_2 + \ldots P_j X_j$$

$$= \sum_{i=1}^{j} P_i X_i$$

where

$$\sum_{i=1}^{j} P_i X_i = 1.0$$

As an example, consider the (X, P) data in Table 8.1.
The expected value of the above data is

$$E(X) = (20)(1/20) + (5)(1/4) + (0)(1/4) - (10)(1/4) - 10(1/5)$$

$$= 1.0 + 1.25 + 0.0 - 2.25 - 2.0$$

$$= 2.0$$

TABLE 8.1
Random Variable/Probability Data

	1	2	3	4	5
X	20	5	0	−10	−10
P	1/20	1/4	1/4	1/4	1/5

EXV 2: Resistance Expected Value

Three resistors, chosen from batches with average resistance of 10, 20, and 30 ohms, are attached in series. What is the expected value of the resistance of the resulting circuit?

Solution

Because the average of a sum of random variables equals the sum of their expected values, the expected value is

$$E = 10 + 20 + 30 = 60 \text{ ohms}$$

EXV 3: Exponential Distribution Expected Value

Find the expected value of a continuous random variable X with the following exponential probability distribution function

$$f(x) = \mu e^{-\mu x}; \quad x \geq 0$$

Solution

Based on the definition of expected values (Equation 8.2),

$$E(X) = \int_{-\infty}^{\infty} x f(x)\, dx$$

Therefore,

$$E(X) = \int_{0}^{\infty} x \mu e^{-\mu x}\, dx = \frac{1}{\mu}$$

The expected values given for the other distributions reviewed in Section II can be similarly calculated.

EXV 4: Expected Value of X

A random variable X assumes the values 10, 20, and 30 with probabilities 0.2, 0.3, and 0.5, respectively. Calculate the expected value of X.

Solution

Once again, recall that the expected value of a random variable is the average value of the random variable. The pdf of X is specified as follows:

$$f(x) = 0.2; \quad x = 10$$

$$= 0.3; \quad x = 20$$

$$= 0.5; \quad x = 30$$

The pdf of X shows what probabilities are assigned to the possible values of X. Once again, use the following formula to calculate the expected value of X.

$$E(X) = \sum_x xf(x)$$

Substituting gives

$$E(X) = 10(0.2) + 20(0.3) + 30(0.5)$$

$$= 2 + 6 + 15$$

$$= 23$$

EXV 5: Expected Value of X^2

Refer to Problem EXV 4, the expected value of X^2.

Solution

Once again, use the following formula (Equation 8.2) to calculate the expected value of X^2.

$$E[g(X)] = \sum_x g(x)f(x)$$

Replacing $g(x)$ with x^2 and substituting leads to

$$E(X^2) = (10^2)(0.2) + (20^2)(0.3) + (30^2)(0.5)$$

$$= 20 + 120 + 450$$

$$= 590$$

EXV 6: Profit Expectations

Let X be a discrete random variable describing the profit in millions of dollars (MM$), with its associated probability distribution of total revenues as shown in Table 8.2. Calculate the following terms:

1. $E(X)$
2. $E(X^2)$
3. $E(2X + 1)^2$

TABLE 8.2
Probability/Distribution of Revenue

X	−3	6	9
P	1/6	1/2	1/3

Solution

1. The expected value of X is

$$E(X) = -3(1/6) + 6(1/2) + 9(1/3) = 33/6 = 11/2 = 5.5$$

2. The expected value of X^2 is

$$E(X^2) = 9(1/6) + 36(1/2) + 81(1/3) = 279/6 = 73/2 = 36.5$$

3. Employ the laws of expectation for this calculation. See Equation 8.5.

$$E\{(2X + 1)^2\} = E\{4X^2 + 4X + 1\}$$

$$= 4E(X^2) + 4E(X) + 1$$

$$= 4(73/2) + 4(11/2) + 1$$

$$= 169$$

EXV 7: EXPONENTIAL DISTRIBUTION VARIANCE

Find the variance of a continuous random variable X with the following exponential distribution:

$$f(x) = \mu e^{-\mu x}$$

Solution

Based on the definition of $E(X)$,

$$E(X) = \int_{-\infty}^{\infty} xf(x)\,dx = \int_{0}^{\infty} x\mu e^{-\mu x}\,dx = \frac{1}{\mu}$$

In addition,

$$E(X^2) = \int_{-\infty}^{\infty} x^2 f(x)\,dx = \int_{0}^{\infty} x^2\mu e^{-\mu x}\,dx = \frac{2}{\mu^2}$$

Substituting into Equation 8.7 gives

$$\text{Var}(X) = \sigma^2 = E(X^2) - [E(X)]^2 = \frac{2}{\mu^2} - \left(\frac{1}{\mu}\right)^2 = \frac{2}{\mu^2} - \frac{1}{\mu^2}$$

$$= \frac{1}{\mu^2}$$

EXV 8: MEAN AND VARIANCE OF A DIE

A die is loaded so that the probability of any face turning up is directly proportional to the number of dots on the face. Let X denote the outcome of throwing the die once. Find the mean and variance of X.

Solution

The pdf of X is

$$f(x) = cx; \quad x = 1, 2, 3, 4, 5, 6$$

Apply the definition of a pdf to determine c.

$$\sum_1^6 f(x) = \sum_1^6 cx = 1$$

Therefore,

$$c = 1/21$$

and

$$f(x) = (1/21)x$$

The mean of x is

$$E(X) = \sum_1^6 xf(x)$$

$$E(X) = \sum_1^6 x^2(1/21)$$

$$= \frac{1}{21} + \frac{4}{21} + \frac{9}{21} + \frac{16}{21} + \frac{25}{21} + \frac{36}{21}$$

$$= \frac{91}{21} = \frac{13}{3} = 4.33$$

For the variance, first calculate $E(X^2)$:

$$E(X^2) = \sum_1^6 x^2 f(x)$$

$$E(X^2) = \sum_1^6 x^3(1/21)$$

$$= \frac{1}{21} + \frac{8}{21} + \frac{27}{21} + \frac{64}{21} + \frac{125}{21} + \frac{216}{21}$$

$$= \frac{441}{21} = 21$$

Once again, the variance of X is

$$\sigma^2 = E(X^2) - [E(X)]^2$$

Substituting the results obtained gives

$$\sigma^2 = 21 - \left(\frac{13}{3}\right)^2$$

$$= \frac{189}{9} - \frac{169}{9}$$

$$= \frac{20}{9} = 2.22$$

EXV 9: PUMP FAILURE EXPECTED VALUES

The time, X, to failure in hours of a certain pump in a wastewater treatment facility has a pdf specified as follows:

$$f(x) = (1/10)\, e^{-x/10}; \quad x > 0$$

Calculate the expected value of X and of X^2.

Solution

The expected value of X is

$$E(X) = \int_0^\infty x(1/10)e^{-x/10}\, dx = 10$$

The expected value of X^2 is

$$E(X^2) = \int_0^\infty x^2(1/10)e^{-x/10}\, dx = 200$$

The calculation of the variance is left as an exercise for the reader. (Hint: don't take a century to solve the problem.)

EXV 10: EXPECTED VALUE AND VARIANCE OF EARTHQUAKE MEASUREMENTS

Refer to Problem PDF 10 in Chapter 7. Determine the expected value and the variance of X.

Solution

Noting that

$$f(x) = 1.7e^{-1.7x}; \quad x > 0$$

$$= 0; \qquad \text{elsewhere}$$

One proceeds as follows:

$$E(X) = \int_0^\infty x(1.7e^{-1.7x})\,dx$$

$$= 0.5882$$

and

$$E(X^2) = \int_0^\infty x^2(1.7e^{-1.7x})\,dx$$

$$= 0.6920$$

The variance of X is then given by

$$\sigma^2 = E(X^2) - [E(X)]^2$$

$$= 0.6920 - (0.5882)^2$$

$$= 0.6920 - 0.3460$$

$$= 0.3460$$

9 EMV. Estimation of Mean and Variance

INTRODUCTION

As discussed earlier, the mean μ and the variance σ^2 of a random variable are constants characterizing the random variable's average value and dispersion about its mean. The mean and variance can also be derived from the probability distribution function (pdf) of the random variable. If the pdf is unknown, however, the mean and the variance can be estimated on the basis of a random sample of some, but not all, observations on the random variable.

Let $X_1, X_2, ..., X_n$ denote a random sample of n observations on X. Then the sample mean $\overline{X}$ is defined by

$$\overline{X} = \sum_{i=1}^{n} \frac{X_i}{n} \tag{9.1}$$

and the sample variance s^2 is defined by

$$s^2 = \sum_{i=1}^{n} \frac{(X_i - \overline{X})^2}{n-1} \tag{9.2}$$

where $\overline{X}$ and s^2 are random variables in the sense that their values vary from the sample to sample of observations on X. It can be shown that the expected value of $\overline{X}$ is μ and that the expected value of s^2 is σ^2. Because of this, $\overline{X}$ and s^2 are called unbiased estimators of μ and σ^2, respectively.

The calculation of s^2 can be facilitated by use of the computation formula

$$s^2 = \frac{n \sum_{i=1}^{n} X_i^2 - \left(\sum_{i=1}^{n} X_i\right)^2}{n(n-1)} \tag{9.3}$$

For example, given the sample 5, 3, 6, 4, 7,

$$\sum_{i=1}^{5} X_i^2 = 135$$

$$\sum_{i=1}^{5} X_i = 25$$

$$n = 5$$

Substituting in Equation 9.3 yields

$$s^2 = \frac{(5)(135)-(25)^2}{(5)(4)}$$

$$= 2.5$$

In the case of a random sample of observations on a continuous random variable assumed to have a so-called normal pdf, the graph of which is a bell-shaped curve, the following statements give a more precise interpretation of the sample standard deviation s as a measure of spread or dispersion.

1. $\overline{X} \pm s$ includes approximately 68% of the sample observations.
2. $\overline{X} \pm 2s$ includes approximately 95% of the sample observations.
3. $\overline{X} \pm 3s$ includes approximately 99.7% of the sample observations.

The source of these percentages is the normal probability distribution, which is studied in more detail in Chapter 18 of Section II.

Chebyshev's theorem provides an interpretation of the sample standard deviation, the positive square root of the sample variance, as a measure of the spread (dispersion) of sample observations about their mean. Chebyshev's theorem states that with $k > 1$, at least $(1 - 1/k^2)$, of the sample observations lie in the interval X $(ks, X + ks)$. For $k = 2$, for example, this means that at least 75% of the sample observations lie in the interval $X(2s, \overline{X_o} + 2s)$. The smaller the value of s, the greater the concentration of observations in the vicinity of X.

PROBLEMS AND SOLUTIONS

EMV 1: DEFINITIONS OF MEASURES OF CENTRAL TENDENCIES

Define the following "central tendency" terms in equation form:

1. Arithmetic mean
2. Geometric mean
3. Median
4. Mode

Solution

Let $X_1, X_2, \ldots, X_n$ represent a set of n numbers.

1. For the arithmetic mean $\overline{X}$,

$$\overline{X} = \frac{X_1 + X_2 + \ldots + X_n}{n} \tag{9.4}$$

If the numbers X_i have weighing factors W_i associated with them,

$$\overline{X} = \frac{W_1 X_1 + W_2 X_2 + \ldots + W_n X_n}{n}$$

$$= \frac{\sum_{i=1}^{n} W_i X_i}{n} \tag{9.5}$$

Weighing factors are often normalized, i.e., $\Sigma W_i = 1.0$. For this condition,

$$\overline{X} = \sum_{i=1}^{n} W_i X_i \tag{9.6}$$

2. The geometric mean, $\overline{X}_G$ is given by

$$\overline{X}_G = (X_1 X_2, ... X_n)^{1/n}$$

$$\overline{X}_G = \left(\prod_{i=1}^{n} x_i \right)^{1/n} \tag{9.7}$$

3. The median is defined as the middle value (or arithmetic mean of the two middle values) of a set of numbers. Thus, the median of 4, 5, 9, 10, 15 is 9. It is also occasionally defined as the distribution midpoint. Further, the median of a continuous probability distribution function $f(x)$ is that value of c so that

$$\int_{-\infty}^{c} f(x)\, dx = 0.5 \tag{9.8}$$

4. The mode is the value that occurs with the greatest frequency in a set of numbers. Thus, it is the typical or most common value in a set.

Other definitions also exist in the literature for the aforementioned four terms.

EMV 2: QUALITATIVE DESCRIPTION OF THE MEAN

Qualitatively describe the mean.

Solution

One basic way of summarizing data is by the computation of a central value. The most commonly used central value statistic is the arithmetic average, or the mean discussed in the previous problem. This statistic is particularly useful when applied to a set of data having a fairly symmetrical distribution. The mean is an efficient statistic because it summarizes all the data in the set and each piece of data is taken into account in its computation. However, the arithmetic mean is not a perfect measure of the true central value of a given data set because arithmetic means can overemphasize the importance of one or two extreme data points.

When a distribution of data is asymmetrical, it is sometimes desirable to compute a different measure of central value. This second measure, known as the *median*, is simply the middle value of a distribution, or the quantity above which half the data lie and below which the other half lie. If n data points are listed in their order of magnitude, the median is the $[(n + 1)/2]$th value. If the number of data is even, then the numerical value of the median is the value midway between the two data points nearest the middle. The median, being a positional value, is less influenced by extreme values in a distribution than the mean. However, the median alone is usually not a good measure of central tendency. To obtain the median, the data provided must first be arranged in order of magnitude, such as

8, 10, 13, 15, 18, 22

Thus, the median is 14, or the value halfway between 13 and 15 because this data set has an even number of measurements.

Another measure of central tendency used in specialized applications is the aforementioned *geometric mean*, X_G.

Generally, the mean falls near the "middle" of the distribution. Actually, the mean may be thought of as the *center of gravity* of the distribution. The mean has another important property. If each measurement is subtracted from the mean, one obtains n "discrepancies" or differences; some of these are positive and some are negative, but the algebraic sum of all the differences is equal to zero.

EMV 3: QUALITATIVE DESCRIPTION OF THE STANDARD DEVIATION

Qualitatively describe the standard deviation.

Solution

The mean of a set of measurements provides some information about the location of the "middle" or "center of gravity" of the set of measurements, but it gives no information about the *scatter* (or *dispersion* or amount of concentration) of the measurements. For example, the five measurements 14.0, 24.5, 25.0, 25.5, and 36.0 have the same mean as the five measurements 24.0, 24.5, 25.0, 25.5, and 26.0, but the two sets of measurements have different amounts of scatter.

One simple indication of the scatter of a set of measurements is the *range*, i.e., the largest measurement minus the smallest. In the two sets of measurements mentioned, the ranges are 22 and 2, respectively. With fairly small sample sizes one would find the range to be very convenient. It is difficult, however, to compare a range for one sample size with that for a different sample size. For this and other reasons, the range, in spite of its simplicity, convenience, and importance, is used only in rather restricted situations.

One clearly needs a measure of scatter, which can be used in samples of any size and in some sense makes use of all the measurements in the sample. There are several measures of scatter that can be used for this purpose, and the most common of these is the *standard deviation*. The standard deviation may be thought of as the "natural" measure of scatter.

Calculation details for the variance and deviation are provided in this chapter and Chapter 8. The examples that follow will illustrate the procedure to obtain these values.

EMV 4: MEDIAN, MEAN, AND STANDARD DEVIATION OF WASTEWATER TEMPERATURES

The average weekly wastewater temperatures (°C) for six consecutive weeks are

$$22, 10, 8, 15, 13, 18$$

Find the median, the arithmetic mean, the geometric mean, and the standard deviation.

Solution

As noted in Problem EMV 2, the median is 14, or the value halfway between 13 and 15, because this data set has an even number of measurements.

For the above wastewater temperatures (substituting T for X),

$$\overline{T}_G = [(8)(10)(13)(15)(18)(22)]^{1/6} = 13.54°C$$

and the arithmetic mean, $\overline{T}$, is

$$\overline{T} = \frac{(8+10+13+15+18+22)}{6}$$

$$= 14.33°C$$

As noted earlier, the most commonly used measure of dispersion, or variability, of sets of data is the standard deviation, s. Its defining formula is given by the expression in Equation 9.2:

$$s = \sqrt{\frac{\sum(X_i - \overline{X})^2}{n-1}}$$

The expression $(X_i - \overline{X})$ shows that the deviation of each piece of data from the mean is taken into account by the standard deviation. Although the defining formula for the standard deviation gives insight into its meaning, the following algebraically equivalent formula (see Equation 9.3) makes computation much easier (now applied to the temperature, T):

$$s = \sqrt{\frac{\sum(T_i - \overline{T})^2}{n-1}} = \sqrt{\frac{n\sum T_i^2 - \left(\sum T_i\right)^2}{n(n-1)}}$$

The standard deviation may be calculated for the data at hand:

$$\sum T_i^2 = (8)^2 + (10)^2 + (13)^2 + (15)^2 + (18)^2 + (22)^2 = 1366$$

and

$$(\sum T_i)^2 = (8 + 10 + 13 + 15 + 18 + 22)^2 = 7396$$

Thus,

$$s = \sqrt{\frac{6(1366) - 7396}{(6)(5)}} = 5.16°C$$

EMV 5: Total Suspended Particulates (TSP) Data

The following are TSP readings ($\mu g/m^3$) for April, May, and June of 2006 at Hallandale, FL — home of the Gulfstream Park racetrack and the Florida Derby (which one of the authors regularly visits):

156 52 138 26 36 66 67 63

Calculate the following terms:

1. Mean
2. Median

3. Mode
4. Range
5. Standard deviation
6. Variance

Solution

1. The mean TSP reading, denoted by $\overline{X}$, is

$$\overline{X} = \frac{\sum_{i=1}^{8} X_i}{n}; \quad n = 8$$

$$\overline{X} = \frac{156 + 52 + 138 + 26 + 36 + 66 + 67 + 63}{8}$$

$$= 75.5 \ \mu g/m^3$$

2. The median, the middle value, is 64.5 $\mu g/m^3$ (63 + 66/2):

$$26 \quad 36 \quad 52 \quad 63 \mid 66 \quad 67 \quad 138 \quad 156$$

Therefore, the median is

$$\frac{63 + 66}{2} = 64.5$$

3. Because the mode is the value of the observation that appears most frequently (if the variable is discrete), there appears to be no mode in this group.
4. The range is given by the maximum value minus the minimum value. Thus, the range, R, is

$$R = 156 - 26 = 130 \ \mu g/m^3$$

5. For the standard deviation, apply Equation 9.2.

$$s = \sqrt{\frac{\sum X_i^2 - \frac{\left(\sum X_i\right)^2}{n}}{n-1}}$$

The calculations are performed in Table 9.1.
From the calculation of $\sum X_i$ determine $(\sum X_i)^2$:

$$(\sum X_i)^2 = (604)^2 = 364{,}816$$

TABLE 9.1
TSP Calculations

X_i	X_i^2
26	676
36	1296
52	2704
63	3969
66	4356
67	4489
138	19044
156	24366
$\Sigma X_i = 604$	$\Sigma X_i^2 = 60870$

Substituting gives

$$s = \sqrt{\frac{60,870 - \dfrac{364,816}{8}}{7}}$$

$$= \sqrt{2181.14}$$

$$= 46.70 \ \mu g/m^3$$

6. By definition, the variance is given as

$$s^2 = (\text{standard deviation})^2$$

$$= (46.70)^2$$

$$= 2181 \ (\mu g/m^3)^2$$

EMV 6: SULFUR DIOXIDE (SO$_2$) DATA

The following are SO$_2$ concentrations ($\mu g/m^3$) for March, April, and May of 2006 at Floral Park, NY — home of the Belmont Park racetrack (which one of the authors regularly visits):

$$29 \ \ 103 \ \ 27 \ \ 14 \ \ 24 \ \ 63 \ \ 24$$

Calculate the following terms:

1. Mean
2. Median
3. Mode
4. Range
5. Standard deviation
6. Variance

Solution

In a very real sense, this is a repeat of Problem EMV 5.

1. For the mean,

$$\overline{X} = \frac{\sum_{i=1}^{8} x_i}{n}; \quad n = 7$$

$$\overline{X} = \frac{29 + 103 + 27 + 14 + 24 + 63 + 24}{7}$$

$$= 40.57 \ \mu g/m^3$$

2. For the median, the middle value is 27 $\mu g/m^3$ since,

$$14 \ \ 24 \ \ 24 \ \ \textcircled{27} \ \ 29 \ \ 63 \ \ 103$$

3. For the mode, the most frequently occurring value is 24 $\mu g/m^3$.
4. The range is given by the maximum value minus the minimum value. Thus, the range, R, is

$$R = 103 - 14 = 89 \ \mu g/m^3$$

5. For the standard deviation, apply Equation 9.3.

$$s = \sqrt{\frac{\sum X_i^2 - \dfrac{\left(\sum X_i\right)^2}{n}}{n-1}}$$

The calculations are performed as follows based on Table 9.2.

TABLE 9.2
SO$_2$ Calculations

X_i	X_i^2
14	196
24	576
24	576
27	729
29	841
63	3969
103	10609
$\sum X_i = 284$	$\sum X_i^2 = 17496$

From the calculation of $\sum X_i$ determine $(\sum X_i)^2$:

$$(\sum X_i)^2 = (284)^2 = 80{,}656$$

Substituting gives

$$s = \sqrt{\dfrac{17{,}496 - \dfrac{80{,}656}{7}}{6}}$$

$$= \sqrt{995.62}$$

$$= 31.55 \; \mu g/m^3$$

6. Finally, the variance is

$$s^2 = (\text{standard deviation})^2$$

$$= (31.55)^2$$

$$= 995 \; (\mu g/m^3)^2$$

EMV 7: CALCULATION AND INTERPRETATION OF MEAN AND VARIANCE: DISCRETE CASE

A discrete random variable X assumes the values 49, 50, and 51 each with probability 1/3. A discrete random variable Y assumes the values 0, 50, and 100 each with probability 1/3. Compare the means and variances of X and Y, and interpret the results.

Solution

As indicated in Chapter 7, the pdf of a discrete random variable requires the probability assigned to each of the possible values of the random variable. For this problem,

$$f(x) = 1/3; \quad x = 49, 50, 51$$

$$g(y) = 1/3; \quad y = 0, 50, 100$$

To obtain μ_x, the mean of X, and μ_y, the mean of Y, use the formulas:

$$\mu_x = E(X) = \sum x \, f(x)$$

$$\mu_y = E(Y) = \sum y \, g(y)$$

Substituting gives,

$$\mu_x = 49(1/3) + 50(1/3) + 51(1/3)$$

$$= 50$$

$$\mu_y = 0(1/3) + 50(1/3) + 100(1/3)$$

$$= 50$$

Obtain σ_x^2, the variance of X, and σ_y^2, the variance of Y, by applying Equation 8.6.

$$\sigma_x^2 = E(X - \mu_x)^2$$

$$\sigma_y^2 = E(X - \mu_y)^2$$

The alternative formula for computing the variance of X is

$$E(X \mu_x)^2 = E(X^2) \mu_x^2$$

Applying this form of the equation to both X and Y gives

$$\sigma_x^2 = \sum (x - \mu_x)^2 f(x)$$

$$= \sum (x - 50)^2 f(x)$$

$$= (49 - 50)^2 (1/3) + (50 - 50)^2 (1/3) + (51 - 50)^2 (1/3)$$

$$= \frac{2}{3} = 0.667$$

$$\sigma_y^2 = \sum (y - \mu_y)^2 g(y)$$

$$= \sum (y - 50)^2 g(y)$$

$$= (0 - 50)^2 (1/3) + (50 - 50)^2 (1/3) + (100 - 50)^2 (1/3)$$

$$= \frac{5000}{3} = 1667$$

Note that X and Y have the same mean but the variance of Y is greater than the variance of X, which reflects the greater dispersion (spread or variability) of the values of Y about its mean in contrast to the dispersion of the values of X about its mean.

EMV 8: CALCULATION AND INTERPRETATION OF MEAN AND VARIANCE: CONTINUOUS CASE

Continuous random variables X and Y have pdfs specified by $f(x)$ and $g(y)$, respectively, as follows:

$$f(x) = 1/2; \quad 1 < x < 1$$

$$g(y) = 1/4; \quad 2 < y < 2$$

Compute the mean and variance of X and Y and compare the results.

Solution

Note that for a continuous random variable, integration replaces summation in the calculation of the mean and variance. Furthermore, the probability that a continuous random variable lies in a certain interval is obtained by integrating the pdf over that interval.

First compute μ_x, the mean of X, and μ_y, the mean of Y. Use the following formulas:

$$\mu_x = \int_{-\infty}^{\infty} x f(x)\,dx$$

$$\mu_y = \int_{-\infty}^{\infty} y g(y)\,dy$$

Substituting $f(x)$ and $g(y)$ gives

$$\mu_x = \int_{-1}^{1} x(1/2)\,dx = 0$$

$$\mu_y = \int_{-2}^{2} y(1/4)\,dy = 0$$

The terms σ_x^2, the variance of X, and σ_y^2, the variance of Y, are calculated as follows:

$$\sigma^2 = E(X^2) - \mu_x^2$$

$$= \int_{-1}^{1} x^2(1/2)\,dx = 1/3 = 0.333$$

$$\sigma^2 = E(Y^2) - \mu_y^2$$

$$= \int_{-2}^{2} y^2(1/4)\,dy = 4/3 = 1.333$$

As with the previous problem, X and Y have the same mean, 0. The variance of Y is greater than the variance of X, which reflects the greater dispersion of the values of Y about its mean.

EMV 9: PROBABILITY DISTRIBUTION FUNCTION: MEAN AND VARIANCE

A continuous random variable X has a pdf given by

$$f(x) = 4x^3; \quad 0 < x < 1$$

Calculate μ and σ^2 for x.

Solution

By definition,

$$\mu = \int x f(x)\,dx$$

Substituting gives

$$\mu = \int_0^1 x(4x^3)\,dx = \int_0^1 4x^4\,dx$$

Integrating gives

$$\mu = 4\left(\frac{x^5}{5}\right)\Bigg|_0^1$$

$$= \frac{4}{5}$$

For σ^2,

$$\sigma^2 = E(X^2) - \mu^2$$

$$= \int x^2 f(x)\,dx - \mu^2$$

Substituting gives

$$\sigma^2 = \int_0^1 x^2(4x^3)\,dx - \mu^2$$

Integrating,

$$\sigma^2 = \frac{4x^6}{6} - \left(\frac{4}{5}\right)^2$$

$$= \frac{2}{3} - \frac{16}{25} = \frac{50 - 48}{75}$$

$$= \frac{2}{75} = 0.0267$$

EMV 10: ESTIMATION OF POPULATION MEAN AND VARIANCE

Estimate the unknown population mean μ and population variance σ^2 on the basis of the following random sample:

$$1, 6, 2, 4, 7$$

Solution

As expected values, the population mean μ and population variance σ^2 can be calculated from the pdf of the random variable when the pdf is known. However, the sample mean is a random variable in the sense that its values vary from sample to sample in the same population.

First, estimate the population mean μ from the sample mean $\overline{X}$, defined as follows for a random sample $X_1, X_2, \ldots, X_n$:

$$\overline{X} = \sum_{i=1}^{n} X_i / n$$

For the given sample, $n = 5$ and $\sum X_i = 20$. Therefore,

$$\overline{X} = 20/5 = 4$$

The sample variance, similar to the population variance, is a measure of dispersion; i.e., the sample variance measures the dispersion of the sample observations about their mean. Similar to the sample mean, the sample variance is a random variable in the sense that it varies from sample to sample in the same population.

Estimate the population variance σ^2 from the sample variance s^2 defined as follows:

$$s^2 = \frac{\sum (X_i - \overline{X})^2}{n - 1}$$

For the given sample, $\overline{X} = 4$ and $n = 5$. Therefore,

$$s^2 = [(1 - 4)^2 + (6 - 4)^2 + (2 - 4)^2 + (4 - 4)^2 + (7 - 4)^2]/4$$

$$= 26/4 = 6.5$$

Section II

Traditional Applications

10 BAY. Bayes' Theorem

INTRODUCTION

Consider n mutually exclusive events $A_1, A_2, ..., A_n$ whose union is the sample space S. Let B be any given event. Then Bayes' theorem states

$$P(A_i\backslash B) = \frac{P(A_i)P(B \backslash A_i)}{\sum_{i=1}^{n} P(A_i)P(B \backslash A_i)}; \quad i = 1, 2, ..., n \tag{10.1}$$

where $P(A_1), P(A_2), ..., P(A_n)$ are called the *prior probabilities* of $A_1, A_2, ..., A_n$ and $P(A_1|B), P(A_2|B), ..., P(A_n|B)$ are called the *posterior probabilities* of $A_1, A_2, ..., A_n$. Also note that the sum of all the posterior probabilities must equal 1.0. This theorem can also be applied to continuous variables, although the applications are rare. Bayes' theorem provides the mechanism for revising prior probabilities, i.e., for converting them into posterior probabilities on the basis of the observed occurrence of some given event.

Suppose, for example, that an explosion at a chemical plant could have occurred as a result of one of three mutually exclusive causes: equipment malfunction, carelessness, or sabotage. It is estimated that such an explosion could occur with a probability of 0.20 as a result of equipment malfunction, 0.40 as a result of carelessness, and 0.75 as a result of sabotage. It is also estimated that the prior probabilities of the three possible causes of the explosion are, respectively, 0.50, 0.35, and 0.15. Using Bayes' theorem, one can now determine the most likely cause of the explosion.

Let A_1, A_2, and A_3 denote, respectively, the events that equipment malfunction, carelessness, and sabotage occur. Let B denote the event of the explosion. Then

$$P(A_1) = 0.50; \quad P(B\backslash A_1) = 0.20$$

$$P(A_2) = 0.35; \quad P(B\backslash A_2) = 0.40$$

$$P(A_3) = 0.15; \quad P(B\backslash A_3) = 0.75$$

Substituting in Equation 10.1, one obtains

$$P(A_1 \backslash B) = \frac{P(A_1)P(B \backslash A_1)}{P(A_1)P(B \backslash A_1) + P(A_2)P(B \backslash A_2) + P(A_3)P(B \backslash A_3)}$$

$$= \frac{(0.50)(0.20)}{(0.50)(0.20) + (0.35)(0.40) + (0.15)(0.75)}$$

$$= 0.28$$

Similarly,

$$P(A_2 \backslash B) = 0.40; \quad P(A_3 \backslash B) = 0.32$$

Therefore, carelessness is the most likely cause of the explosion.

As another example of the application of Bayes' theorem, suppose that 50% of a company's particle size analyzer output comes from a New York plant, 30% from a Pennsylvania plant, and 20% from a Delaware plant. On the basis of plant records, it is estimated that defective analyzers constitute 1% of the output of the New York plant, 3% of the Pennsylvania plant, and 4% of the Delaware plant. If an analyzer selected at random from the company's manufactured output is found to be defective, what are the revised probabilities that the analyzer was produced by each of the three plants?

Let A_1, A_2, A_3 denote, respectively, the events that the analyzer was produced in the New York, Pennsylvania, and Delaware plants. Let B denote the event that the analyzer was found to be defective. Then

$$P(A_1) = 0.50; \quad P(B \backslash A_1) = 0.01$$

$$P(A_2) = 0.30; \quad P(B \backslash A_2) = 0.03$$

$$P(A_3) = 0.20; \quad P(B \backslash A_3) = 0.04$$

Substituting in Equation 10.1 gives

$$P(A_1 \backslash B) = \frac{P(A_1)P(B \backslash A_1)}{P(A_1)P(B \backslash A_1) + P(A_2)P(B \backslash A_2) + P(A_3)P(B \backslash A_3)}$$

$$= \frac{(0.50)(0.01)}{(0.50)(0.01) + (0.30)(0.03) + (0.20)(0.04)}$$

$$= 0.23$$

Similarly,

$$P(A_2 \backslash B) = 0.41; \quad P(A_3 \backslash B) = 0.36$$

Therefore the information that the analyzer selected at random was defective "revises" the probability that the analyzer was produced in the New York plant downward from 0.50 to 0.23 — i.e., 23% of all defective devices come from New York — and revises the probabilities for Pennsylvania and Delaware upward, respectively, from 0.30 to 0.41 and from 0.20 to 0.36.

For another example of the use of Bayes' theorem, suppose that the probability is 0.80 that an airplane crash due to structural failure is diagnosed correctly. Suppose, in addition, that the probability is 0.30 that an airplane crash due to something else, but not structural failure, is incorrectly attributed to structural failure. If 35% of all airplane crashes are due to structural failure, what is the probability that an airplane crash was due to structural failure, given that it has been so diagnosed?

Let A_1 be the event that structural failure is the cause of the airplane crash. Let A_2 be the event that the cause is other than structural failure. Let B be the event that the airplane crash is diagnosed as being due to structural failure. Then,

$$P(A_1) = 0.35; \quad P(B \backslash A_1) = 0.80$$

$$P(A_2) = 0.65; \quad P(B \backslash A_2) = 0.30$$

Substituting in Equation 10.1 gives

$$P(A_1 \backslash B) = \frac{P(A_1)P(B \backslash A_1)}{P(A_1)P(B \backslash A_1) + P(A_2)P(B \backslash A_2)}$$

$$= \frac{(0.35)(0.80)}{(0.35)(0.80) + (0.65)(0.30)}$$

$$= 0.59$$

Therefore, the diagnosis of structural failure revises its probability as the cause of the airplane crash upward from 0.35 to 0.59.

PROBLEMS AND SOLUTIONS

BAY 1: STATISTICIAN'S DILEMMA

Discuss the two major "interpretations" of Bayes' theorem.

Solution

In many applications, *a priori* probabilities are obtained from supposed knowledge. However, in many situations, the initial knowledge is not that well defined. Statisticians of the subjectivist school who think of probability in terms of "degree of belief" recommend the use of Bayes' theorem even when *a priori* estimates are based on personal judgment, such as the evaluation by a marketing manager of the probability that a product will "sell," or an engineer's estimate based on past experience with similar systems, that a particular missile will "fly." In fact, this school of statisticians has been referred to as Bayesians because it makes very frequent use of Bayes' theorem in its work. This is in contrast to the majority of statisticians who, though accepting Bayes' theorem as a consequence of the axioms of probability, feel that — based upon their interpretation of probability — no reasonable estimate of *a priori* probabilities is available in most applied situations.

BAY 2: PRODUCTION LINE QUERY

A production line uses two processes; 80% of the production lots emanate from process 1 and 20% from process 2. It is known from past experience that process 1 and process 2 yield 5 and 25% bad units, respectively. A very large lot, presumably selected at random, is available, and one must decide whether the lot comes from process 1 or from process 2. If one unit randomly selected from the lot is found to be defective, estimate the posterior probabilities of the lot emanating from process 1 and process 2.

Solution

Let B_1 represent the event that the lot comes from process 1, and let B_2 represent the event that the lot comes from process 2. Initially, $P(B_1) = 0.80$ and $P(B_2) = 0.20$ because on average, four out of every five randomly chosen lots do come from process 1. Assume that one unit randomly selected from the lot is found to be defective. Designate this outcome as A. Now the probabilities that a random unit from a lot from process 1 and process 2 would be defective are 0.05 and 0.25, respectively; i.e.,

$$P(A \backslash B_1) = 0.05; \quad P(A \backslash B_2) = 0.25$$

One can then combine the *a priori* knowledge with the result from the sample and estimate the posterior probabilities of the lot emanating from process 1 and process 2. From Equation 10.1,

$$P(B_1 \backslash A) = \frac{P(B_1)P(A \backslash B_1)}{P(B_1)P(A \backslash B_1) + P(B_2)P(A \backslash B_2)}$$

Substituting gives

$$P(B_1 \backslash A) = \frac{(0.80)(0.05)}{(0.80)(0.05) + (0.20)(0.25)}$$

$$= 0.444$$

$$P(B_2 \backslash A) = \frac{(0.20)(0.25)}{(0.80)(0.05) + (0.20)(0.25)}$$

$$= 0.556$$

Note that the consequence of the defective sample has been to change the *a priori* probability from odds of 4 to 1 in favor of the lot emanating from process 1 [$P(B_1) = 0.80$] to odds of 5 to 4 [i.e., $P(B_1|A) = 0.444$] in favor of the lot emanating from process 2. Once again, the 0.444 result only means that 44% of all defective items come from process 1; 80% of all units are still made by process 1.

BAY 3: EFFECT OF SELECTION PROCESS

Refer to Problem BAY 2. Instead of picking a defective unit, a nondefective unit has been selected. Calculate the revised posterior probabilities.

Solution

Designate the selection of a nondefective unit as the outcome A. Then,

$$P(\overline{A}|B_1) = 0.95; \quad P(\overline{A}\backslash B_2) = 0.75$$

and

$$P(B_1|\overline{A}) = \frac{(0.80)(0.95)}{(0.80)(0.95) + (0.20)(0.75)}$$

$$= 0.835$$

$$P(B_2|\overline{A}) = \frac{(0.20)(0.75)}{(0.80)(0.95) + (0.20)(0.75)}$$

$$= 0.165$$

In this case, the sample result tended to "confirm" the *a priori* probability. The consequence is to change the probability that the lot has emanated from process 1 only slightly — from odds of 4 to 1 in favor $[P(B_1|A) = 0.535]$ to about 5 to 1.

BAY 4: PLANT ORIGIN OF CONTAMINATED DRUGS

Suppose that 50% of a pharmaceutical company's manufactured drug comes from plant 1, 30% from plant 2, and 20% from plant 3. On the basis of plant records, it is estimated that contaminated drugs constitute 0.1% of the output of plant 1, 0.3% of the output of plant 2, and 0.4% of the output of plant 3. If a drug is selected at random from the company's manufactured output and is found to be contaminated, what are the revised probabilities that the item was produced by each of the three plants?

Solution

Represent the given event by B; i.e., B is the event that the drug selected at random was defective. Identify the antecedent events and represent them by $A_1, A_2, ..., A_n$. Thus, A_1 is the event that the drug was selected from plant 1; A_2, plant 2; A_3, plant 3. The prior probabilities are

$$P(B \backslash A_1) = 0.001$$

$$P(B \backslash A_2) = 0.003$$

$$P(B \backslash A_3) = 0.004$$

One can now substitute these values in Equation 10.1 (Bayes' theorem) to determine the revised (posterior) probabilities, i.e., $P(A_1|B), P(A_2|B), ..., P(A_n|B)$:

$$P(A_1 \backslash B) = \frac{(0.50)(0.001)}{(0.50)(0.001)+(0.30)(0.003)+(0.20)(0.004)}$$

$$= \frac{0.0005}{0.0005+0.0009+0.0008}$$

$$= \frac{0.0005}{0.0022}$$

$$\doteq 0.23$$

Similarly,

$$P(A_2 \backslash B) = 0.41$$

$$P(A_3 \backslash B) = 0.36$$

Note that, as expected, the sum of the revised probabilities equals unity.

BAY 5: VOLTAGE REGULATORS

An environmental instrumentation company received 75% of its voltage regulators from supplier A and 25% from supplier B. It was found that 90% of the regulators from A and 85% of the regulators from B perform according to specifications. What is the probability that a regulator came from supplier A, given that it performs according to specifications?

Solution

Let A_1 be the event that the voltage regulator came from supplier A and let A_2 be the event that the voltage regulator came from supplier B. Let B be the event that the voltage regulator performs according to specifications. Therefore,

$$P(A_1) = 0.75; \quad P(A_2) = 0.25$$

$$P(B\backslash A_1) = 0.90; \quad P(B\backslash A_2) = 0.85$$

Application of Bayes' theorem yields

$$P(A_1\backslash B) = \frac{(0.75)(0.90)}{(0.75)(0.90)+(0.25)(0.85)}$$

$$= 0.76$$

BAY 6: CAUSE OF A NANOCHEMICAL PLANT EXPLOSION

An explosion at a nanochemical plant could have occurred as a result of three mutually exclusive causes involving nanoparticles: particle size distribution (PSD), particle concentration (PC), and particle shape (PS). It is estimated that such an explosion could occur with a probability of 0.75 as a result of PSD, 0.40 as a result of PC, and 0.30 as a result of PS. It is also estimated that the prior probabilities of the three possible causes are, respectively, 0.60, 0.30, and 0.10. What is the most likely cause of the explosion?

Solution

Represent the given event by B; i.e., B is the explosion of the nanoparticles. Identify the antecedent events and represent them by $A_1, A_2, \ldots, A_n$. Let A_1, A_2, A_3 denote, respectively, the events that PSD, PC, and PS are the causes of the explosion. The prior probabilities are

$$P(A_1) = 0.60$$

$$P(A_2) = 0.30$$

$$P(A_3) = 0.10$$

The conditional probabilities of the event B given each of the antecedent events are

$$P(B\backslash A_1) = 0.75$$

$$P(B\backslash A_2) = 0.40$$

$$P(B\backslash A_3) = 0.30$$

Once again, substitute these values in Equation 10.1 to determine the revised (posterior) probabilities, i.e.,

$$P(A_1\backslash B), P(A_2\backslash B), \ldots, P(A_n\backslash B)$$

$$P(A_1 \backslash B) = \frac{(0.60)(0.75)}{(0.60)(0.75)+(0.30)(0.40)+(0.10)(0.30)}$$

$$= \frac{0.45}{0.45+0.12+0.03}$$

$$= 0.75$$

$$P(A_2\backslash B) = 0.20$$

$$P(A_3\backslash B) = 0.05$$

A_1 has the highest posterior probability. Therefore, PSD is the most likely cause of the explosion. The occurrence of the explosion results in the revision of the prior probability of PSD upward from 0.60 to 0.75, PC downward from 0.30 to 0.20, and PS downward from 0.10 to 0.05.

BAY 7: HEALTH INSURANCE COMPANY

A health insurance company has high-, medium-, and low-risk policyholders who have probabilities 0.02, 0.01, and 0.0025, respectively, of filing a claim within any given year. The proportions of company policyholders in the three risk groups are 0.10, 0.20, and 0.70. What proportion of claims filed each year come from the low-risk group?

Solution

Let A_1 be the event that a policyholder is at high risk, A_2 be the event that a policyholder is at medium risk, and A_3 be the event that a policyholder is at low risk. Let B be the event that a policyholder files a claim. Based on the problem statement,

$$P(A_1) = 0.10; \quad P(A_2) = 0.20; \quad P(A_3) = 0.70$$

$$P(B\backslash A_1) = 0.02; \quad P(B\backslash A_2) = 0.01; \quad P(B\backslash A_3) = 0.0025$$

Application of Bayes' theorem yields

$$P(B\backslash A_3) = \frac{(0.70)(0.0025)}{(0.70)(0.0025)+(0.20)(0.01)+(0.10)(0.02)}$$

$$= 0.30$$

BAY 8: SMALL-DIAMETER CYCLONES

A company manufactures small-diameter cyclones (for installation in multi-cyclones employed for air pollution control) in three plants, say, A, B, and C. On the average, 4 cyclones out of 500 from A, 10 out of 800 from B, and 10 out of 1000 from C must be recalled. If a customer purchases a cyclone from a dealer and it is recalled, what is the probability of the following events, if 30, 40, and 30 of the dealer's cyclones come from plants A, B, and C, respectively:

1. The cyclone came from plant A.
2. It came from plant B.
3. It came from plant C.

Solution

Let A, B, and C represent the events that a cyclone is delivered from plant A, plant B, and plant C, respectively. If D is the event of a recall, then by Bayes' theorem, the probabilities are as follows:

1. Event 1:

$$P(A \backslash D) = \frac{P(D \backslash A)P(A)}{P(D \backslash A)P(A) + P(D \backslash B)P(B) + P(D \backslash A)P(C)}$$

$$= \frac{(4/500)(3/10)}{(4/500)(3/10) + (10/800)(4/10) + (10/1000)(3/10)}$$

$$= 0.2308$$

2. Event 2:

$$P(B \backslash D) = \frac{P(D \backslash B)P(B)}{P(D)}$$

$$= \frac{(10/800)(4/10)}{(0.0104)}$$

$$= 0.4808$$

3. Event 3:

$$P(C \backslash D) = \frac{P(D \backslash C)P(C)}{P(D)}$$

$$= \frac{(10/1000)(3/10)}{(0.0104)}$$

$$= 0.2884$$

As a check, observe that the three probabilities sum to 1.0 (unity).

11 SPS. Series and Parallel Systems

INTRODUCTION

Many systems consisting of several components can be classified as *series*, *parallel*, or a combination of both. However, the majority of industrial and process plants (units and systems) have series of parallel configurations.

A *series system* is one in which the entire system fails to operate if any one of its components fails to operate. If such a system consists of n components that function independently, then the reliability of the system is the product of the reliabilities of the individual components. If R_s denotes the reliability of a series system and R_i denotes the reliability of the ith component $i = 1, \ldots, n$, then

$$R_s = R_1 R_2 \ldots R_n$$
$$= \prod_{i=1}^{n} R_i \tag{11.1}$$

A *parallel system* is one that fails to operate only if all its components fail to operate. If R_i is the reliability of the ith component, then $(1 - R_i)$ is the probability that the ith component fails; $i = 1, \ldots, n$. Assuming that all n components function independently, the probability that all n components fail is $(1 - R_1)(1 - R_2) \ldots (1 - R_n)$. Subtracting this product from unity yields the following formula for R_p, the reliability of a parallel system.

$$R_p = 1 - (1 - R_1)(1 - R_2) \ldots (1 - R_n)$$
$$= \prod_{i=1}^{n} (1 - R_i) \tag{11.2}$$

The reliability formulas for series and parallel systems can be used to obtain the reliability of a system that combines features of a series and a parallel system as shown in Figure 11.1. These calculations are illustrated in this chapter.

PROBLEMS AND SOLUTIONS

SPS 1: COMBINED SERIES AND PARALLEL SYSTEMS

Consider the system diagrammed in Figure 11.2. Components A, B, C, and D have for their respective reliabilities 0.90, 0.90, 0.80, and 0.90. The system fails to operate if A fails, if B and C both fail, or if D fails. Determine the reliability of the system.

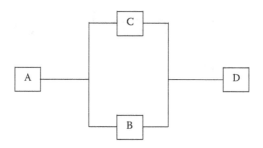

FIGURE 11.1 System with parallel and series components.

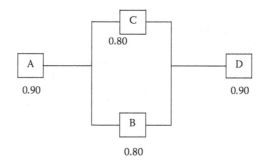

FIGURE 11.2 System with parallel and series component values.

Solution

Components B and C constitute a parallel subsystem connected in series to components A and D. The reliability of the parallel subsystem is obtained by applying Equation 11.2, which yields

$$R_p = 1 - (1 - 0.80)(1 - 0.80) = 0.96$$

The reliability of the system is then obtained by applying Equation 11.1, which yields

$$R_s = (0.90)(0.96)(0.90) = 0.78$$

SPS 2: RELIABILITY OF AN ELECTRICAL SYSTEM

Determine the reliability of the electrical system shown in Figure 11.3 using the reliabilities indicated under the various components.

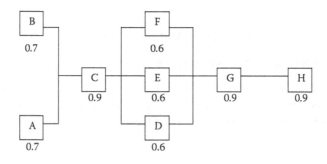

FIGURE 11.3 Diagram of electrical system I.

Solution

First identify the components connected in parallel. A and B are connected in parallel. D, E, and F are also connected in parallel. Then compute the reliability of each subsystem of the components connected in parallel. The reliability of the parallel subsystem consisting of components A and B is

$$R_p = 1 - (1 - 0.7)(1 - 0.7) = 0.91$$

The reliability of the parallel subsystem consisting of components D, E, and F is

$$R_p = 1 - (1 - 0.6)(1 - 0.6)(1 - 0.6) = 0.936$$

Multiply the product of the reliabilities of the parallel subsystems by the product of the reliabilities of the components to which the parallel subsystems are connected in series:

$$R_s = (0.91)(0.9)(0.936)(0.9)(0.9) = 0.621$$

The reliability of the whole system is therefore 0.621.

SPS 3: Reliability of Overseas Flight

A failure analysis of a military overseas flight is regarded as a series system with the following components: ground crew (A), cockpit crew (B), aircraft (C), weather conditions (D), and landing accommodations (E). The cockpit crew is viewed as a parallel system with the following components: captain (B_1), copilot (B_2), and flight engineer (B_3). Landing accommodations are viewed as a parallel system with the following components: scheduled airport (E_1), alternate landing sites (E_2 and E_3). Failure probabilities for the various components are estimated as follows:

$A = 0.001$	$B_3 = 0.100$	$E_1 = 0.001$
$B_1 = 0.001$	$C = 0.001$	$E_2 = 0.050$
$B_2 = 0.010$	$D = 0.0001$	$E_3 = 0.100$

What is the probability of a successful flight?

Solution

First identify the components connected in parallel. B_1, B_2, and B_3 are connected in parallel. E_1, E_2, and E_3 are also connected in parallel. As with Problem SPS 2, compute the reliability of each subsystem of the components connected in parallel. The reliability of the parallel subsystem consisting of components B_1, B_2, and B_3 is

$$R_p = 1 - (1 - 0.999)(1 - 0.99)(1 - 0.90) = 0.999999$$

The reliability of the parallel subsystem consisting of the components E_1, E_2, and E_3 is

$$R_p = 1 - (1 - 0.999)(1 - 0.95)(1 - 0.90) = 0.999995$$

Multiply the product of the reliabilities of the parallel subsystems by the product of the reliabilities of the components to which the parallel subsystems are connected in series:

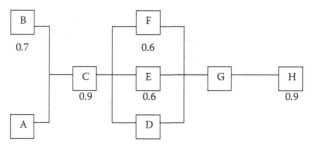

FIGURE 11.4 Diagram of electrical system II.

$$R_s = (0.999999)(0.999995)(0.999)(0.999)(0.9999) = 0.9979$$

The probability of a successful flight is therefore 0.9979 or 99.79%.

SPS 4: UNKNOWN COMPONENT RELIABILITIES I

Determine the reliability of the components A, D, and G of the electrical system illustrated in Figure 11.4. Use the reliabilities indicated under the various components. The overall reliability has been determined to be 0.42.

Solution

The reliability of the parallel subsystem consisting of components A and B is obtained by applying Equation 11.2, which yields

$$R_p = 1 - (1 - 0.7)(1 - A)$$

$$= 1 - (-0.3 - A + 0.7A)$$

$$= 0.7 + 0.3A$$

The reliability of the parallel subsystem consisting of components D, E, and F is

$$R_p = 1 - (1 - D)(1 - 0.6)(1 - 0.6)$$

$$= 1 - 0.16(1 - D)$$

$$= 0.84 + 0.16D$$

The reliability of the entire system is obtained by applying Equation 11.1, which yields

$$R_s = 0.42 = (0.7 + 0.3A)(0.9)(0.84 + 0.16D)(G)(0.9)$$

This single equation contains three unknowns. An infinite number of solutions are possible, including, for example,

$$A = 0; \quad D = 0; \quad G = 0.882$$

SPS 5: UNKNOWN COMPONENT RELIABILITIES II

Refer to Problem SPS 4. Determine the reliability of component D if $A = 0.7$.

Solution

With $A = 0.7$, the equation for R_s (Equation 11.1) becomes

$$R_s = 0.42 = (0.91)(0.9)(0.84 + 0.16D)(G)(0.9)$$

There is one equation and two unknowns. Once again, an infinite number of solutions are possible, including, for example,

$$D = 0; \quad G = 0.678$$

SPS 6: UNKNOWN COMPONENT RELIABILITIES III

Refer to Problem SPS 4 and Problem SPS 5. Determine the reliability of component G if $A = 0.7$ and $D = 0.09$.

Solution

Here,

$$R_s = 0.42 = (0.91)(0.9)(0.84 + 0.16(0.09))(G)(0.9)$$

$$G = 0.667$$

SPS 7: EXPONENTIAL FAILURE RATE: SERIES SYSTEM

Consider the system shown in Figure 11.5. Determine the reliability, R, if the operating time for each unit is 5000 h. Components A and B have exponential failure rates, λ, of 3×10^{-6} and 4×10^{-6} failures per hour, respectively, where $R_i = e^{-\lambda_i t}$; t = time, h. The term λ may be viewed as the reciprocal of the average time to failure. See Chapter 16 for additional details.

Solution

Because this is a series system,

$$R_s = R_A R_B$$

As indicated earlier, for an exponential failure rate

$$R = e^{-\lambda t}; \quad t = \text{time}, h$$

so that

$$R_A = e^{(3E-6)(5000)} = e^{0.015} = 0.9851$$

FIGURE 11.5 Exponential failure rate: series system.

and

$$R_B = e^{(4E-6)(5000)} = e^{0.02} = 0.9802$$

Therefore,

$$R_s = (0.9851)(0.9802) = 0.9656$$

SPS 8: Series Components Reversed

Refer to Problem SPS 7. Recalculate the reliability of the system if the order is reversed, i.e., A follows B.

Solution

Because

$$R_s = R_A R_B = R_B R_A$$

the reliability remains the same.

SPS 9: Exponential Failure Rate: Parallel System

Consider the system shown in Figure 11.6. Determine the reliability of this system employing the information provided in Problem SPS 7.

Solution

Because this is a parallel system,

$$R_p = 1 - (1 - R_A)(1 - R_B)$$

Employing the results from Problem SPS 8,

$$R_p = 1 - (1 - 0.9851)(1 - 0.9802)$$

$$= 1.0 - (0.0149)(0.0198)$$

$$= 0.9997$$

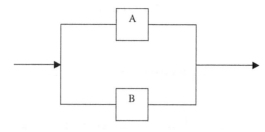

FIGURE 11.6 Exponential failure rate: parallel system.

SPS 10: PARALLEL COMPONENTS INTERCHANGED

Refer to Problem SPS 9. Recalculate the reliability if the two components are interchanged.

Solution

Because

$$R_p = 1 - (1 - R_A)(1 - R_B)$$

$$= 1 - (1 - R_B)(1 - R_A)$$

the reliability again remains the same.

12 BIN. The Binomial Distribution

INTRODUCTION

Consider n independent performances of a random experiment with mutually exclusive outcomes that can be classified as *success* or *failure*. The words "success" and "failure" are to be regarded as labels for two mutually exclusive categories of outcomes of the random experiment. They do not necessarily have the ordinary connotation of success or failure. Assume that p, the probability of success on any performance of the random experiment, is constant. Let Q be the probability of failure, so that

$$Q = 1 - p \tag{12.1}$$

The probability distribution of X, the number of successes in n performances of the random experiment, is the binomial distribution, with a probability distribution function (pdf) specified by

$$f(x) = \frac{n!}{x!(n-x)!} p^x Q^{n-x}; \quad x = 0, 1, \ldots, n \tag{12.2}$$

where $f(x)$ is the probability of x successes in n performances. One can show that the expected value of the random variable X is np and its variance is npQ.

As a simple example of the binomial distribution, consider the probability distribution of the number of defectives in a sample of 5 items drawn with replacement from a lot of 1000 items, 50 of which are defective. Associate "success" with drawing a defective item from the lot. Then the result of each drawing can be classified as success (defective item) or failure (nondefective item). The sample of five items is drawn with replacement (i.e., each item in the sample is returned before the next is drawn from the lot; therefore, the probability of success remains constant at 0.05). Substituting in Equation 12.2 the values $n = 5$, $p = 0.05$, and $Q = 0.95$ yields

$$f(x) = \frac{5!}{x!(5-x)!} (0.05)^x (0.95)^{5-x}; \quad x = 0, 1, 2, 3, 4, 5$$

as the pdf for X, the number of defectives in the sample. The probability that the sample contains exactly three defectives is given by

$$P(X = 3) = \frac{5!}{3!2!} (0.05)^3 (0.95)^2 = 0.0011$$

The binomial distribution can be used to calculate the reliability of a *redundant system*. A redundant system consisting of n identical components is a system that fails only if more than r components fail. Familiar examples include single-usage equipment such as missile engines, short-life batteries, and flash bulbs, which are required to operate for one time period and are not

TABLE 12.1
Binomial Probabilities

X	Binomial n = 20, P = 0.05	Binomial n = 100, P = 0.01
0	0.3585	0.3660
1	0.3774	0.3697
2	0.1887	0.1849
3	0.0596	0.0610
4	0.0133	0.0149
5	0.0022	0.0029
6	0.0003	0.0005
≥ 7	0.0000	0.0001

reused. Once again, associate "success" with the failure of a component. Assume that the n components are independent with respect to failure, and that the reliability of each is $1 - p$. Then X, the number of failures, has the binomial pdf in Equation 12.2 and the reliability of the redundant system is

$$P(X \leq r) = \sum_{x=0}^{r} \frac{n!}{x!(n-x)!} p^x Q^{n-x} \tag{12.3}$$

Additional details are provided in Chapter 13.

Tables that can assist binomial distribution calculations are available in the literature. Probabilities for two different binomial coefficients in Equation 12.2 are presented in Table 12.1.

The binomial distribution occurs in problems in which samples are drawn from a large population with specified "success" or "failure" probabilities and there is a desire to evaluate instances of obtaining a certain number of successes in the sample. It has applications in quality control, reliability, environmental management, consumer sampling, and many other cases.

Suppose a trial or event can result in one and only one of mutually exclusive events E_1, E_2, ..., E_j with probabilities p_1, p_2, ..., p_j respectively, where $p_1 + p_2 + \dots + p_j = 1$. If n independent trials are made, then one can show that the probability function of obtaining $x_1 E_1$s, $x_2 E_2$s, ..., $x_j E_j$s is given by

$$f(x_1, \dots, x_j) = \frac{n!}{x_1! \cdots x_j!} p_1^{x_1} \cdots p_j^{x_j} \tag{12.4}$$

where $0 \leq x_j \leq n$, $i = 1, \dots, j$, and $x_1 + \dots + x_j = n$. This is the probability function of the *multinomial distribution*. Note that if $j = 2$ this function reduces to the binomial distribution. Hence, the multinomial distribution is essentially an extension of the binomial distribution for just two possible outcomes — success and failure.

PROBLEMS AND SOLUTIONS

BIN 1: One-Die Application

Provide a binomial distribution example involving one die.

Solution

Consider 100 tosses of a die. Associate "success" with getting a 1 or a 2 on a single toss. Associate "failure" with getting a 3, 4, 5, or 6 on a single toss. The probability of success on a single toss is 1/3. The probability of failure is 2/3. Note that the outcomes of successive tosses are independent of each other. This is an example of 100 performances of a random experiment (tossing a die), with each performance resulting in either success with constant probability 1/3 or failure with constant probability 2/3.

There are many practical applications involving n independent performances of a random experiment such as the one just mentioned. Interest usually centers on the number X of successes. As described in the Introduction section, the random variable X has a binomial probability distribution with pdf specified by Equation 12.2:

$$f(x) = \frac{n!}{x!(n-x)!} p^x Q^{n-x}; \quad x = 0, 1, \ldots, n$$

where $f(x)$ is the probability of x successes in n independent performances of a random experiment resulting in success with $Q = 1 - p$. And further, as described earlier, success and failure do not have the ordinary connotation associated with these words; i.e., they are merely labels designating two mutually exclusive categories for the classification of the outcome of each performance of the random experiment under consideration.

BIN 2: SAMPLING WITH REPLACEMENT I

A sample of five transistors is drawn with replacement from a lot which is 5% defective. Once again, "with replacement" means that each transistor drawn is returned to the lot before the next is drawn. What is the probability that the number of defective transistors in the sample is exactly 2?

Solution

This random experiment consists of drawing a transistor at random with replacement from a lot. The random experiment is performed five times because a sample of five transistors is drawn with replacement from the lot. Therefore, $n = 5$. Also note that the performances are independent because each transistor is replaced before the next is drawn. Therefore, the composition of the lot is exactly the same before each drawing.

For this problem, associate "success" with drawing a defective transistor. Associate "failure" with drawing a nondefective. Refer once again to Equation 12.2. Because 5% of the lot is defective, $p = 0.05$. Therefore, $Q = 0.95$. Substitute these values (n, p, and Q) in the binomial pdf.

$$f(x) = \frac{5!}{x!(5-x)!} (0.05)^x (0.95)^{5-x}; \quad x = 0, 1, \ldots, 5$$

Substitute the appropriate value of X to obtain the required probabilities.

$$P \text{ (exactly 2 defectives)} = \frac{5!}{2!3!} (0.05)^2 (0.95)^3 = 0.214$$

BIN 3: SAMPLING WITH REPLACEMENT II

Refer to Problem BIN 2. What is the probability of the following:

1. The number of defective transistors is at most two.
2. The number of defectives is at least two.

Solution

For event 1,

$$P \text{ (at most 2 defectives)} = \sum_{x=0}^{2} \frac{5!}{x!(5-x)!}(0.05)^x(0.95)^{5-x}$$

$$= 0.7738 + 0.2036 + 0.0214$$

$$= 0.9988$$

For event 2,

$$P \text{ (at least 2 defectives)} = P(X \geq 2)$$

$$= 1 - P(X \leq 1)$$

$$= 1 - \sum_{x=0}^{1} \frac{5!}{x!(5-x)!}(0.05)^x(0.95)^{5-x}$$

$$= 1 - (0.7748 + 0.2036)$$

$$= 0.0226$$

Note that this problem involved summing the binomial pdf over the appropriate values of X; this procedure applies if the probability of more than a single value of X is required.

BIN 4: NANOCARCINOGEN APPLICATION

The probability that an exposure to a nanocarcinogen will be fatal is 0.80. Find the probability of the following events for a group of 15 workers:

1. At least 9 will die.
2. From 4 to 8 will die.

Solution

For event 1,

$$P \text{ (at least 9 will die)} = P(X \geq 9); \quad p = 0.8, Q = 0.2$$

$$= \sum_{x=9}^{15} \frac{15!}{x!(15-x)!}(0.8)^x(0.2)^{15-x}$$

$$= 0.0430 + 0.1032 + 0.1876 + 0.2501 + 0.2309 + 0.1319 + 0.0352$$

$$= 0.982$$

This calculation can be performed by longhand or obtained directly from binomial tables similar to that provided in Table 12.1.

For event 2,

$$P(4 \leq X \leq 8) = 1.0 - P(X \geq 9) - P(0 \leq X \leq 4)$$

One notes almost immediately that

$$P(0 \leq X \leq 4) = 0$$

Therefore,

$$P(4 \leq X \leq 8) = 1.0 - 0.982 - 0.0$$

$$= 0.018$$

BIN 5: MANUFACTURING OPERATION

A plant manufactures filter bags in very large lots by a standard production process. A customer selects a sample of 20 bags at random from each production lot and rejects the complete lot if he or she finds 4 or more bad bags. If, in fact, the production process yields exactly 10% defectives, what is the probability of lot rejection?

Solution

The problem may be rephrased as follows: "Given a binomial process with probability of success 0.1, what is the probability of 4 or more successes in 20 independent trials?" This equals

$$P \text{ (lot rejection)} = \sum_{x=4}^{20} \frac{20!}{x!(20-x)!} (0.1)^x (0.9)^{20-x}$$

$$= 1 - \sum_{x=0}^{3} \frac{20!}{x!(20-x)!} (0.1)^x (0.9)^{20-x}$$

$$= 1 - (0.1216 + 0.2702 + 0.2852 + 0.1901)$$

$$= 1 - 0.8671$$

$$= 0.1329$$

Once again, this calculation can be performed by longhand or obtained directly from binomial tables.

BIN 6: SPRAY SYSTEM

A nanoparticle production unit contains (for cooling purposes) 20 independent sprays, each of which fails with a probability of 0.10. The system fails only if four or more of the sprays fail. What is the probability that the unit will fail?

Solution

Let X denote the number of sprays that fail. The term X has a binomial distribution with $n = 20$ and $p = 0.10$. The probability that the system fails is given by

$$P(X \geq 4) = \sum_{x=4}^{20} \frac{20!}{x!(20-x)!}(0.1)^x(0.9)^{20-x}$$

$$= 1 - \sum_{x=0}^{3} \frac{20!}{x!(20-x)!}(0.1)^x(0.9)^{20-x}$$

$$= 0.1329$$

The reader should note the similarity of this problem with Problem BIN 5.

BIN 7: SYNTHETIC DIAMONDS

An engineer's ability to distinguish a natural from a synthetic nano-produced diamond is tested independently on ten different occasions. What is the probability of seven correct identifications if the engineer is only guessing?

Solution

For guessing purposes, the probability of making a correct identification may be reasonably assumed to be 0.5. Let X denote the number of correct identifications. If the engineer is only guessing, X has a binomial distribution with $n = 10$ and $p = 0.5$. The probability of exactly seven correct identifications is

$$P(X = 7) = \frac{10!}{7!3!}(0.5)^7(0.5)^3$$

$$= 0.1172$$

BIN 8: LOT SAMPLE INSPECTION

A procuring agent for an environmental engineering firm is asked to sample a lot of 100 pumps. The sampling procedure calls for the inspection of 20 pumps. If there are any bad pumps, the lot is rejected; otherwise it is accepted. The chief engineer has asked the agent the following questions:

1. When there are four bad pumps in the lot, how often would one expect to accept the lot?
2. Suppose one bad pump is allowed in the sample. What kind of protection would the company (agent) have?

Solution

In this problem, $n = 20$ and $p = 4/100 = 0.04$.
    For case 1,

$$P(X = 0) = \frac{20!}{0!20!}(0.04)^0(0.96)^{20}$$

$$= 0.442$$

For case 2, one needs to calculate the probability of accepting the lot (with four defectives) if 0 or 1 defective is allowed in the sample. Therefore,

$$P(X \leq 1) = P(X = 0) + P(X = 1)$$

$$= 0.442 + \frac{20!}{1!19!}(0.04)^1(0.96)^{19}$$

$$= 0.442 + (20)(0.04)(0.4604)$$

$$= 0.442 + 0.368$$

$$= 0.810$$

BIN 9: CASINO SIMULCAST BETTING

One of the authors recently bet on ten basketball games at the Mirage simulcasting center in Las Vegas. Assuming the odds of winning the bet are 0.5, what is the probability of breaking even, i.e., winning 5 of the bets.

Solution

For this application, Equation 12.2 applies with $n = 10$, $p = 0.5$, and $X = 5$.

$$P(X = 5) = \frac{10!}{5!5!}(0.5)^5(0.5)^5$$

$$= (252)(0.03125)(0.031525)$$

$$= 0.246$$

BIN 10: NUMBER OF GAMES BET

Refer to Problem BIN 9. How many games would the author have to bet so that the probability of breaking even is

1. 50%
2. 25%

Solution

For case 1, the number of games is given by

$$P(X = n/2) = \frac{2!}{1!1!}(0.5)^n$$

$$0.5 = (2)(0.5)^n$$

$$n = 2 \text{ (as expected)}$$

For case 2, no exact integer number of games can be determined,

$$P(X = n/2) = (252)(0.5)^n$$

$$0.25 = (252)(0.5)^n$$

$$n = 9.98 \text{ (by trial and error)}$$

or alternatively,

$$\frac{0.25}{252} = (0.5)^n$$

$$\log\left(\frac{0.25}{252}\right) = n\log(0.5)$$

$$n = \frac{\log\left(\dfrac{0.25}{252}\right)}{\log(0.5)}$$

$$= 9.98$$

This number when rounded off to the nearest integer is 10, a result that could have been deduced immediately from Problem BIN 9.

BIN 11: RELIABILITY OF A REDUNDANT SYSTEM

A redundant system consisting of three operating pumps can survive two pump failures. Assume that the pumps are independent with respect to failure and each has a probability of failure of 0.10. What is the reliability of the system?

Solution

The system consists of three pumps. Therefore, $n = 3$. The system can survive the failure of two pumps. Equation 12.3 should be employed with $r = 2$. The probability of a pump failure is 0.10. Therefore, $p = 0.10$. Substitute these values of n and p in the binomial pdf:

$$f(x) = \frac{3!}{x!(3-x)!}(0.10)^x(0.90)^{3-x}; \quad x = 0, 1, \ldots, 3$$

Sum the binomial pdf from 0 to r to find the reliability, R:

$$R = \sum_{x=0}^{2}\frac{3!}{x!(3-x)!}(0.10)^x(0.90)^{3-x}$$

$$= 0.999$$

BIN 12: MULTINOMIAL DISTRIBUTION

The probabilities of hospitals A, B, and C obtaining a particular type of serum are 0.5, 0.3, and 0.2, respectively. Four such serums are to be provided. What is the probability that a single hospital receives all four serums?

Solution

This involves the use of the multinomial distribution. The desired probability is (see Equation 12.4)

$$P(1 \text{ hospital alone getting serum}) = f(4, 0, 0) + f(0, 4, 0) + f(0, 0, 4)$$

$$= \frac{4!}{4!0!0!}(0.5)^4(0.3)^0(0.2)^0 + \frac{4!}{0!4!0!}(0.5)^0(0.3)^4(0.2)^0 + \frac{4!}{0!0!4!}(0.5)^0(0.3)^0(0.2)^4$$

$$= 0.0625 + 0.0081 + 0.0016$$

$$= 0.0722$$

Note that for this calculation each hospital receives all four shipments.

13 REL. Reliability Relations

INTRODUCTION

One of the major applications of statistics involves reliability calculations. This Introduction section derives and sets forth some of the important equations employed in both reliability theory and applications. This material will be extended to include some of the major distributions in subsequent chapters of Section II.

The reliability of a component will frequently depend on the length of time it has been in service. Let T, the time to failure, be the random variable having its probability distribution function (pdf) specified by $f(t)$. Then the probability that failure occurs in the time interval (0,t) is given by (see Chapter 7 in Section I)

$$F(t) = \int_0^t f(t)\, dt \tag{13.1}$$

Let the reliability of the component be denoted by R(t), the probability that the component survives to time t. Therefore,

$$R(t) = 1 - F(t) \tag{13.2}$$

Equation 13.2 establishes the relationship between the reliability of a component and the cumulative distribution function (cdf) of its time to failure.

In accordance with the property expressed earlier (see Chapter 7), the probability that a component will fail in the time interval $(t, t + \Delta t)$ is given by

$$P(t < T < t + \Delta t) = F(t + \Delta t) - F(t) \tag{13.3}$$

The conditional probability that a component will fail in the time interval $(t, t + \Delta t)$, given that it has survived to time t, is

$$P[(t < T < (t + \Delta t)/(T > t)] = \frac{F(t + \Delta t) - F(t)}{P(T > t)} \tag{13.4}$$

Equation 13.4 is obtained by application of the definition of conditional probability. Noting that

$$R(t) = P(T > t) \tag{13.5}$$

and substituting in Equation 13.5, leads to

$$P[t < T < t + (\Delta t / (T > t))] = \frac{F(t + \Delta t) - F(t)}{R(t)} \tag{13.6}$$

Division of both sides of Equation 13.6 by Δt yields

$$\frac{P[t < (T < (t)+\Delta t)/(T > t))]}{\Delta t} = \left[\frac{F(t+\Delta t)-F(t)}{\Delta t}\right]\left[\frac{1}{R(t)}\right] \qquad (13.7)$$

Recall that $F'(t)$, the derivative of $F(t)$, is defined by

$$\lim_{\Delta t \to 0}\left[\frac{F(t+\Delta t)-F(t)}{\Delta t}\right] = F'(t) \qquad (13.8)$$

By taking the limit of both sides of Equation 13.8 as Δt approaches 0,

$$Z(t) = \frac{F'(t)}{R(t)} \qquad (13.9)$$

where $Z(t)$ is defined by

$$Z(t) = \lim_{\Delta t \to 0}\frac{P[(t < T < t+\Delta t/(T > t))]}{\Delta t} \qquad (13.10)$$

Here $Z(t)$ is called the *failure rate* (also the *hazard rate*) of the component. Equation 13.10 establishes the relationship between failure rate reliability and the cdf of time to failure.

Using Equation 13.2 and Equation 13.5, one can obtain an expression for the reliability in terms of the failure rate. Differentiating both sides of Equation 13.9 with respect to t yields

$$R'(t) = 0 - F'(t)$$
$$= -F'(t) \qquad (13.11)$$

Substitution in Equation 13.9 yields

$$Z(t) = -\frac{R'(t)}{R(t)} \qquad (13.12)$$

Integrating both sides of Equation 13.12 between 0 and t yields

$$\int_0^t Z(t)\,dt = -[\ln R(t) - \ln R(0)] \qquad (13.13)$$

Because $R(t) = P(T > t)$, and $R(0) = 1$, Equation 13.13 becomes

$$\int_0^t Z(t)\,dt = -\ln R(t) \qquad (13.14)$$

Solving Equation 13.14 for R(t) yields

$$R(t) = \exp\left[-\int_0^t Z(t)\,dt\right] \tag{13.15}$$

the desired expression for the reliability in terms of failure rate.

The pdf of time to failure can also be expressed in terms of failure rate. Differentiating Equation 13.15 with respect to t yields

$$R'(t) = -Z(t)\exp\left[-\int_0^t Z(t)\,dt\right] \tag{13.16}$$

Equation 13.2 may also be written (by differentiating both sides)

$$R'(t) = -f(t) \tag{13.17}$$

Equation 13.6 can therefore be written as

$$f(t) = Z(t)\exp\left[-\int_0^t Z(t)\,dt\right] \tag{13.18}$$

which represents the desired expression for the pdf in terms of failure rate.

Examples illustrating the application of these equations to real systems are provided in subsequent chapters, particularly with exponential, Weibull, and normal distributions. This chapter is limited to exponential distributions. However, in addition to the exponential, Weibull, and normal distributions, several other probability distributions figure prominently in reliability calculations. Their pdfs, principal characteristics, and an indication of their applications are presented in later chapters.

PROBLEMS AND SOLUTIONS

REL 1: HAZARD DEFINITION

Define hazard.

Solution

Hazard, risk, failure, and reliability are interrelated concepts concerned with uncertain events and therefore amenable to quantitative measurement via probability. *Hazard* is generally defined as a potentially dangerous event; e.g., the release of toxic fumes, a power outage, or pump failure. Actualization of the potential danger represented by a hazard results in undesirable consequences associated with risk.

REL 2: RISK DEFINITION

Define risk.

Solution

Risk is defined as the product of two factors: (1) the probability of an undesirable event and (2) the measured consequences of the undesirable event. Measured consequences may be stated in terms of financial loss, injuries, deaths, or other variables. *Failure* represents an inability to perform some required function. *Reliability* is the probability that a system or one of its components will perform its intended function under certain conditions for a specified period. The reliability of a system and its probability of failure are complementary in the sense that the sum of these two probabilities is unity. In this chapter, one considers the basic concepts and theorems of probability that find application in the estimation of risk and reliability.

REL 3: Battery Reliability

A battery employed at an incineration site is deemed reliable if it operates for more than 500 h. The lives of the previous 11 batteries, in hours, were

$$501, 591, 621, 386, 942, 503, 201, 1013, 902, 32, 899$$

Estimate the reliability of a battery employed at that site.

Solution

Assuming all batteries come from the same population, one notes that 3 of the 11 did not function beyond 500 h. Therefore, the reliability of the battery is simply given by:

$$R = \frac{8}{11}$$

$$= 0.727$$

$$= 72.7\%$$

REL 4: Batteries in Series

Refer to Problem REL 3. If two batteries in series are required for a retrofitted unit, estimate the reliability of the two batteries.

Solution

Refer to Equation 11.1 in Chapter 11. Because the two batteries, with $R = 0.727$, are connected in series,

$$R_s = R_1 R_2$$

$$= (0.727)(0.727)$$

$$= 0.529$$

$$= 52.9\%$$

REL 5: Batteries in Parallel

Refer to Problem REL 3 and Problem REL 4. Resolve Problem REL 4 if the two batteries are in a parallel system.

Solution

For a parallel system, Equation 11.2 in Chapter 11 applies.

$$R_p = 1 - (1 - R_1)(1 - R_2)$$

$$= 1 - (1 - 0.727)(1 - 0.727)$$

$$= 0.924$$

$$= 92.4\%$$

REL 6: REQUIRED RELIABILITY

If a pumping system in a nuclear power plant must have a reliability of 99.99%, how many pumps are required in a parallel system if each pump has a reliability of 95.1%?

Solution

Refer to Equation 11.2 in Chapter 11. The required reliability for this parallel system is

$$R_p = 1 - (1 - R_1)(1 - R_2), \ldots, (1 - R_n)$$

where λR_i is the fractional reliability of pump i.

If the reliabilities of the pumps are assumed equal, i.e.,

$$R_1 = R_2 = \ldots = R_n = R$$

then

$$R_p = 1 - (1 - R)^n$$

Noting that

$$R_p = 0.9999; \quad R = 0.951$$

The solution to the equation yields

$$n = 3$$

REL 7: BATTERY LIFE: EXPONENTIAL MODEL

Employing the information provided in Problem REL 3 and assuming battery life can be reasonably described by an exponential equation, estimate the reliability that a battery would survive 100 h.

Solution

The average time to failure, t_f, is

$$t_f = \frac{(501 + 591 + 621 + 386 + 942 + 503 + 201 + 1013 + 902 + 32 + 899)}{11}$$

$$= \frac{6591}{11}$$

$$= 599 \text{ h}$$

Refer to Problem SPS 7 in Chapter 11. For an exponential model,

$$R = e^{-\lambda t}$$

where λ is the reciprocal of the average time to failure. Thus,

$$\lambda = \frac{1}{t_f} = \frac{1}{599} = 0.00167$$

The reliability for 100 h is:

$$R = e^{-(0.00167)(100)}$$

$$= e^{-0.167}$$

$$= 0.846$$

$$= 84.6\%$$

The reader may refer to Chapter 16 for additional details on exponential models.

REL 8: Adjusted Reliability Time

Refer to Problem REL 7. Recalculate the battery reliability for

1. 500 h
2. 1000 h

Solution

The same reliability equation is employed with t set to the time as required in this problem.

1. For this case, $t = 500$ h, so that

$$R = e^{-(0.00167)(500)}$$

$$= 0.434$$

$$= 43.4\%$$

2. For case 2, $t = 1000$ h, so that

$$R = e^{-(0.00167)(1000)}$$

$$= 0.188$$

$$= 18.8\%$$

REL 9: Reliability Time Comparison

Comment on the results of Problem REL 8.

Solution

As expected as t increases, the reliability decreases, i.e.,

$$R(500) = 0.434$$

$$R(1000) = 0.188$$

REL 10: RELIABILITY TIME ANOMALY

Explain the different results obtained in Problem REL 3 and Problem REL 8 (Part 1).

Solution

For Problem REL 3, the reliability of the battery was 72.7%, whereas the reliability of the battery in Problem REL 8 (Part 1) was 43.4%. This difference arises because the distribution of failure times given in Problem REL 3 does not come from an exponential distribution, i.e., it does not come from a population that has a constant failure rate.

14 HYP. Hypergeometric Distribution

INTRODUCTION

The hypergeometric distribution is applicable to situations in which a random sample of r items is drawn without replacement from a set of n items. *Without replacement* means that an item is not returned to the set after it is drawn. Recall that the binomial distribution is frequently applicable in cases where the item is drawn *with replacement*.

Suppose that it is possible to classify each of the n items as a *success* or *failure*. Again, the words "success" and "failure" do not have the usual connotation. They are merely labels for two mutually exclusive categories into which n items have been classified. Thus, each element of the population may be dichotomized as belonging to one of two disjointed classes.

Let a be the number of items in the category labeled *success*. Then $n - a$ will be the number of items in the category labeled *failure*. Let X denote the number of successes in a random sample of r items drawn without replacement from the set of n items. Then the random variable X has a hypergeometric distribution whose probability distribution function (pdf) is specified as follows:

$$f(x) = \frac{\dfrac{a!}{x!(a-x)!} \dfrac{(n-a)!}{(r-x)!(n-a-r+x)!}}{\dfrac{n!}{r!(n-r)!}}; \quad x = 0, 1, \ldots, \min(a, r) \tag{14.1}$$

The term $f(x)$ is the probability of x successes in a random sample of n items drawn without replacement from a set of n items, a of which are classified as successes and $n - a$ as failures. The term $\min(a,r)$ represents the smaller of the two numbers a and r, i.e., $\min(a,r) = a$ if $a < r$ and $\min(a,r) = r$ if $r \le a$.

The hypergeometric distribution can also be arrived at through the use of combinations (see Chapter 1 in Section I). Using this approach, the total number of possible selections of r elements out of n is $C(n,r)$. Furthermore, x good elements may be chosen out of the total of a good elements in a total of $C(a,x)$ combinations. For each such combination, it is also possible to select $r - x$ of $n - a$ bad elements in $C(a,x)\, C(n-a, r-x)$. Because each selection possibility is equally likely, the probability of picking exactly x good elements is

$$f(x) = \frac{C(a, x)C(n - a, r - x)}{C(n, r)} \tag{14.2}$$

which, upon substitution of C (see Chapter 7) is the expression given in Equation 14.1 for the probability density function of the hypergeometric distribution. As with the binomial distribution, a tabulation of cumulative and individual terms of the hypergeometric distribution is available in the literature.

The hypergeometric distribution is applicable in situations similar to those when the binomial density is used, except that samples are taken from a small population. Examples arise in sampling from small numbers of chemical, medical, and environmental samples, as well as from manufacturing lots.

PROBLEMS AND SOLUTIONS

HYP 1: Relationships between Hypergeometric and Binomial Distributions

Discuss the difference between the hypergeometric and binomial distributions.

Solution

The hypergeometric distribution is obviously a special case of the binomial distribution when applied to finite populations. In particular, the hypergeometric distribution approaches the binomial distribution when the population size approaches infinity. Note, however, that others have claimed that the binomial distribution is a special case of a hypergeometric distribution.

HYP 2: Sampling Transistors without Replacement

A sample of 5 transistors is drawn at random without replacement from a lot of 1000 transistors, 50 of which are defective. What is the probability that the sample contains exactly 3 defectives?

Solution

The number of items in the set from which the sample is drawn is the number of transistors in the lot. Therefore, $n = 1000$. Associate "success" with drawing a defective transistor, and "failure" with drawing a nondefective one. Determine a, the number of "successes" in the set of n items. Because 50 of the transistors in the lot are defective, $a = 50$. Also note that the sample is drawn without replacement and that the size of the sample is $r = 5$.

Substituting the values of n, r, and a in the hypergeometric pdf provided in Equation 14.1 gives

$$f(x) = \frac{\dfrac{50!}{x!(50-x)!}\dfrac{950!}{(5-x)!(945+x)!}}{\dfrac{1000!}{5!995!}}; \quad x = 0, 1, \ldots, 5$$

Also, substitute the appropriate value of X above to obtain the required probability.

$$P(\text{sample contains exactly 3 defectives}) = P(X = 3)$$

Therefore,

$$P(X=3) = \frac{\dfrac{50!}{x!47!}\dfrac{950!}{2!(945+3)!}}{\dfrac{1000!}{5!995!}}$$

$$= 0.0011$$

HYP 3: Pillbox with Drugs

A pillbox contains 24 drug tablets, 3 of which are contaminated. If a sample of six is chosen at random from the pillbox, what is the probability that zero will be contaminated?

Solution

First note that this involves sampling without replacement. Employing the combination equation provided in the introductory section of this chapter, Equation 14.2 yields

$$f(x) = \frac{C(a, x)C(n - a, r - x)}{C(n, r)}; \quad a = 3,\ r = 6,\ n = 24$$

Therefore,

$$f(x) = \frac{C(3, x)C(21, 6 - x)}{C(24, 6)}; \quad 0 \le x \le 3$$

There are only four values X can assume: 0, 1, 2, and 3. Substituting gives

$$P(X = 0) = \frac{C(3,0)C(21,6)}{C(24,6)}$$

$$= \frac{\left(\dfrac{3!}{0!\,3!}\right)\left(\dfrac{21!}{6!\,15!}\right)}{\left(\dfrac{24!}{6!\,18!}\right)}$$

$$= \frac{(1)(54,264)}{(134,596)}$$

$$= 0.4032$$

HYP 4: Additional Contaminated Pills

Refer to Problem HYP 3. Resolve the problem if the number of contaminated pills in the sample of six is

1. One
2. Two
3. Three

Solution

1. For one pill,

$$P(X = 1) = \frac{C(3,1)C(21,5)}{C(24,6)}$$

$$= 0.45356$$

2. For two pills,

$$P(X=2) = \frac{C(3,2)C(21,4)}{C(24,6)}$$

$$= 0.13340$$

3. For three pills,

$$P(X=3) = \frac{C(3,3)C(21,3)}{C(24,6)}$$

$$= 0.00988$$

Note that $\sum_{x=0}^{3} P(X) = 1$.

HYP 5: DIFFERENT SAMPLE SIZES I

Refer to Problem HYP 3. Resolve the problem if the sample size chosen is 8, rather than 6.

Solution

The only difference in this problem is that $r = 8$. The describing equation therefore becomes

$$P(X=0) = \frac{C(3,0)C(21,8)}{C(24,8)}$$

$$= \frac{\left(\dfrac{3!}{0!\,3!}\right)\left(\dfrac{1!}{8!\,13!}\right)}{\left(\dfrac{24!}{8!\,16!}\right)}$$

$$= \frac{(1)(203,490)}{(735,471)}$$

$$= 0.277$$

$$= 27.7\%$$

As expected, the probability decreases.

HYP 6: DIFFERENT SAMPLE SIZES II

Refer to Problem HYP 5. Resolve the problem if the sample size chosen is 2 rather than 8.

Solution

The describing hypergeometric equation now becomes

$$P(X=0) = \frac{C(3,0)\,C(21,2)}{C(24,2)}$$

$$= \frac{(1)(210)}{276}$$

$$= 0.761$$

The binomial equation is

$$P(X=0) = (0.875)^2$$

$$= 0.765$$

As expected, the agreement is excellent.

HYP 7: ACCEPTANCE SAMPLING PLAN EVALUATION I

A quality control engineer inspects a random sample of 3 vacuum pumps drawn without replacement from each incoming lot of 25 vacuum pumps. A lot is accepted only if the sample contains no defectives. If any defectives are found, the entire lot is inspected, and the cost charged to the vendor. What is the probability of accepting a lot containing 10 defective vacuum pumps?

Solution

Each lot consists of 25 vacuum pumps. Therefore, $n = 25$. Once again, associate "success" with a defective vacuum pump. Determine, a, the number of "successes" in the set of n items. For this problem, $a = 10$. A sample of 3 vacuum pumps is drawn without replacement from each lot; therefore, $r = 3$.

Substitute the values of n, r, and a in the pdf of the hypergeometric distribution.

$$f(x) = \frac{\dfrac{10!}{x!(10-x)!}\dfrac{15!}{(3-x)!(12+x)!}}{\dfrac{25!}{3!22!}}; \quad x = 0, 1, 2, 3$$

Substitute the appropriate values of X to obtain the required probability. The probability of accepting a lot containing ten defective vacuum pumps is

$$P(X=0) = \frac{\dfrac{10!}{0!10!}\dfrac{15!}{(3-0)!12!}}{\dfrac{25!}{3!22!}}$$

$$= \frac{(1)(455)}{(2300)}$$

$$= 0.20$$

HYP 8: Acceptance Sampling Plan Evaluation II

Refer to Problem HYP 7. What is the probability of accepting an entire lot containing only one defective vacuum pump?

Solution

The pdf with only one defective pump is

$$f(x) = \frac{\dfrac{1!}{x!(1-x)!} \dfrac{24!}{(3-x)!(21+x)!}}{\dfrac{25!}{3!22!}}; \quad x = 0, 1$$

For no defective vacuum pumps,

$$n = 25, a = 1, r = 3, x = 0$$

Therefore,

$$P(X=0) = \frac{\dfrac{1!}{0!1!} \dfrac{24!}{3!21!}}{\dfrac{25!}{3!22!}}$$

$$= \frac{\dfrac{24!}{3!21!}}{\dfrac{25!}{3!22!}}$$

$$= \frac{2024}{2300}$$

$$= 0.88$$

HYP 9: Water Pollution Monitor Testing

An order to manufacture a special type of water pollution monitor was recently received. The total order consists of 25 monitors. Five of these monitors are to be selected at random and initially life-tested. The contract specifies that if not more than 1 of the 5 monitors fails a specified life test, the remaining 20 will be accepted. Otherwise, the complete lot will be rejected. What is the probability of lot acceptance if exactly 4 of the 25 submitted monitors are defective?

Solution

The lot will be accepted if the sample of 5 contains either 0 or 1 of the 4 defective monitors. The probability of this happening is given by the hypergeometric distribution as

$$P(X \leq 1) = \frac{C(4, x)C(21, 5 - x)}{C(24, 5)}$$

$$= P(X = 0) + P(X = 1)$$

$$= \frac{\frac{4!}{4!0!}\frac{21!}{5!16!}}{\frac{25!}{20!5!}} + \frac{\frac{4!}{3!1!}\frac{21!}{4!17!}}{\frac{25!}{20!5!}}$$

$$= \frac{(1)(20349)}{(53130)} + \frac{(4)(5985)}{(53130)}$$

$$= \frac{20349}{53130} + \frac{23940}{53130}$$

$$= 0.383 + 0.451$$

$$= 0.834$$

HYP 10: SAMPLING NUMBER CHANGE

Refer to Problem HYP 9. Outline how the calculation would be affected if six monitors are selected for life-testing.

Solution

The hypergeometric distribution again applies with the r term now equal to 6 (not 5). Therefore,

$$P(X \leq 1) = \sum_{X=0}^{1} \frac{C(4, x)C(21, 6 - x)}{C(25, 6)}$$

Solving this problem is left as an exercise for the reader.

15 POI. Poisson Distribution

INTRODUCTION

The probability distribution function (pdf) of the Poisson distribution can be derived by taking the limit of the binomial pdf as $n \to \infty$, $P \to 0$, and $nP = \mu$ remains constant. The Poisson pdf is given by

$$f(x) = \frac{e^{-\mu}\mu^x}{x!}; \quad x = 0, 1, 2, \ldots \tag{15.1}$$

Here, $f(x)$ is the probability of x occurrences of an event that occurs on the average μ times per unit of space or time. Both the mean and the variance of a random variable X having a Poisson distribution are μ.

The Poisson pdf can be used to approximate probabilities obtained from the binomial pdf given earlier when n is large and p is small. In general, good approximations will result when n exceeds 10 and P is less than 0.05. When n exceeds 100 and nP is less than 10, the approximation will generally be excellent. Table 15.1 compares binomial and Poisson probabilities for the case of $n = 20$ ($P = 0.05$) and $n = 100$ ($P = 0.01$).

If λ is the failure rate (per unit of time) of each component of a system, then λt is the average number of failures for a given unit of time. The probability of x failures in the specified unit of time is obtained by substituting $\mu = \lambda t$ in Equation 15.1 to obtain

$$f(x) = \frac{e^{-\lambda t}(\lambda t)^x}{x!}; \quad x = 0, 1, 2, \ldots \tag{15.2}$$

Suppose, for example, that in a certain country the average number of airplane crashes per year is 2.5. What is the probability of 4 or more crashes during the next year? Substituting $\lambda = 2.5$ and $t = 1$ in Equation 15.2 yields

$$f(x) = \frac{e^{-2.5}(2.5)^x}{x!}; \quad x = 0, 1, 2, \ldots$$

as the pdf of X, the number of airplane crashes in 1 year. The probability of 4 or more airplane crashes next year is then

$$P(X \geq 4) = 1 - \sum_{x=0}^{3} \frac{e^{-2.5}(2.5)^x}{x!}$$

$$= 1 - (0.0821 + 0.205 + 0.257 + 0.214)$$

$$= 1 - 0.76$$

$$= 0.24$$

TABLE 15.1
Binomial/Poisson Comparison

x	Binomial ($n = 20$, $P = 0.05$)	Poisson ($n = 20$, $P = 0.05$)	Binomial ($n = 100$, $P = 0.01$)	Poisson ($n = 100$, $P = 0.01$)
0	0.3585	0.3679	0.3660	0.3679
1	0.3774	0.3679	0.3697	0.3679
2	0.1887	0.1839	0.1849	0.1839
3	0.0596	0.0613	0.0610	0.0613
4	0.0133	0.0153	0.0149	0.0153
5	0.0022	0.0031	0.0029	0.0031
6	0.0003	0.0005	0.0005	0.0005
≥ 7	0.0000	0.0001	0.0001	0.0001

As another example, suppose that the average number of breakdowns of personal computers during 1000 h of operation of a computer center is 3. What is the probability of no breakdowns during a 10-h work period? Note that the given average is the average number of breakdowns during 1000 h. The unit of time associated with the given average is 1000 h. The probability required is the probability of no breakdowns during a 10-h period, and the unit of time connected with the required probability is 10 h. If 10 is divided by 1000, the result is 0.01 so that in a 10-h period there are 0.01 time periods of 1000-h duration. The given average of breakdown is 3. Multiplication by 0.01 yields 0.03, the average number of occurrences during a 10-h time period, μ. This value of μ may be substituted in the Poisson pdf:

$$f(x) = \frac{e^{-0.03}(0.03)^x}{x!}; \quad x = 0, 1, 2, \ldots$$

Substitute for x, the number of occurrences whose probability is required. The probability of no breakdowns in a 10-h period is:

$$P(X = 0) = \frac{e^{-0.03}(0.03)^0}{0!} = e^{-0.03} = 0.97$$

In addition to the applications cited, the Poisson distribution can be used to obtain the reliability R of a standby redundancy system in which one unit is in the operating mode and n identical units are in standby mode. Unlike parallel systems, the standby units are inactive. The reliability of the standby redundancy system may be calculated by employing the Poisson distribution under the following conditions: (1) all units have the same failure rate in the operating mode, (2) unit failures are independent, (3) standby units have zero failure rate in the standby mode, and (4) there is perfect switchover to a standby when the operating unit fails. This is treated later in this chapter in Problem POI 9.

The Poisson distribution is a distribution in its own right, which arises in many other different situations. For instance, it provides probabilities of specified numbers of telephone calls per unit interval of time, of environmental sampling procedures, of given numbers of defects per unit area of glass, textiles, or papers, and of various numbers of bacterial colonies per unit volume, etc.

PROBLEMS AND SOLUTIONS

POI 1: Computer Breakdowns

The New York Racing Organization (NYRO) has reported that the average number of times a race-horse breaks down and has to be euthanized during the running of 10,000 races is 30. What is the probability that no horses breakdown over a 10-race period?

Solution

As with the example in the introductory section of this chapter, the unit of "time" associated with the probability to be calculated is 10 races. If 10 is divided by 10,000 the result is 0.001, i.e., the 10-race period represents 0.001 races of a total of 10,000 races. Because the average number of horses breaking down in the 10,000-race period is 30, multiplication of 30 by 0.001 yields 0.03. This represents μ, the average number of breakdowns during a 10-race period. Substituting this into Equation 15.2 yields

$$f(x) = \frac{e^{-0.03}(0.03)^x}{x!}; \quad x = 0, 1, 2, \dots$$

as the pdf of X, the number of breakdowns in a 10-race period. The probability of no breakdowns in a 10-race period is then

$$P(X = 0) = e^{-0.03} = 0.97$$

The reader should determine why this result is identical to that calculated in the Introduction section of this chapter.

POI 2: Microscopic Slides

Microscopic slides of a certain culture of microorganisms contain on the average 20 microorganisms per square centimeter. After treatment by a chemical, 1 cm^2 is found to contain only 10 such microorganisms. If the treatment had no effect, what would be the probability of finding ten or fewer microorganisms in a given square centimeter?

Solution

Let X denote the number of microorganisms in 1 cm^2. If the chemical treatment had no effect, X has a Poisson pdf with $\mu = 20$. The probability of ten or fewer microorganisms in a given square centimeter is

$$P(X \leq 10) = \sum_{x=0}^{10} e^{-20}(20)^x$$

$$= P(X = 0) + P(X = 1) + \dots + P(X = 10)$$

Longhand calculation leads to

$$P(X \leq 10) = 0.0128$$

For example,

$$P(X = 0) = 2 \times 10^{-9} \text{ (as expected)}$$

$$P(X = 1) = 4.1 \times 10^{-8}$$

$$P(X = 9) = 0.00291$$

$$P(X = 10) = 0.0058$$

POI 3: Defective Welds

The average number of defective welds detected at the final examination of the tail section of an aircraft is 5. What is the probability of detecting at least one defective weld during the final examination of the tail section?

Solution

The given average is the average number of defective welds detected at the final inspection of an aircraft tail section. The unit of space connected with the given average is the space occupied by the tail section of the aircraft. Therefore, the associated unit of space is the same as the space occupied by the tail section of the aircraft. Because both the unit of space connected with the given average and the unit of space connected with the required area are the same, their quotient is one. The probability required is the probability of detecting at least one defective weld at the final inspection of the tail section. The given average number of defectives is five. Multiplication by one yields the value of μ as five. Substitute this value of μ in the Poisson pdf.

$$f(x) = \frac{e^{-5}(5)^x}{x!}; \quad x = 0, 1, 2, \ldots$$

Substitute for x, the number of occurrences whose probability is required. The probability of detecting at least one defective weld is

$$P(X \geq 0) = 1 - P(X = 0)$$

$$= 1 - \frac{e^{-5}(5)^0}{0!}$$

$$= 0.9933$$

POI 4: Defective Welds Increase

Refer to Problem POI 3. Discuss how an increase in the number of defective welds would be handled.

Solution

The Poisson pdf gives the probability of x occurrences over the unit of space associated with the required probability. As indicated earlier, one must substitute for x in the pdf the number of occurrences whose probability is required. However, more than one value of x may have to be substituted if the required probability is the sum of the probabilities for several values of x.

POI 5: Probability of Dying

The probability that U.S. citizens of age 72 to 73 (the age of one of the authors) will die within the year is 0.0417. With a group of 100 such individuals, what is the probability that exactly 5 will die within the year?

Solution

For this problem,

$$P = 0.0417$$

$$n = 100$$

$$\mu = Pn = (100)(0.0417)$$

$$= 4.17$$

$$r = 5$$

$$P(5) = \frac{(4.17)^5}{5!\,e^{4.17}}$$

$$= \frac{1260}{(120)(64.7)}$$

$$= 0.162$$

POI 6: Hospital Deaths

Over the last 10 years, a local hospital reported that the number of deaths per year due to temperature inversions (air pollution) was 0.5. What is the probability of exactly three deaths in a given year?

Solution

For this problem,

$$P(X = 3) = \frac{e^{-0.5}(0.5)^3}{3!}$$

$$= 0.0126$$

POI 7: Increase in Hospital Deaths

Refer to Problem POI 6. Calculate the annual probability of three or more deaths being attributed to temperature inversions.

Solution

For this case,

$$P(X \geq 3) = \sum_{x=3}^{\infty} \frac{e^{-0.5}(0.5)^x}{x!}$$

$$= 1 - \sum_{x=0}^{2} \frac{e^{-0.5}(0.5)^x}{x!}$$

$$1 - 0.60065 - 0.3033 - 0.0758$$

$$= 0.0227$$

POI 8: UNDEFEATED KENTUCKY DERBY HORSES

In the 155 years that the Kentucky Derby has been run, only 68 horses that entered the race have been undefeated. What is the probability that two undefeated horses will be entered in the next "Run for the Roses"?

Solution

For this problem,

$$\mu = \frac{68}{155}$$

$$= 0.439$$

$$n = 2$$

$$P(2) = \frac{e^{-0.439}(0.439)^2}{2!}$$

$$= 0.062$$

In effect, approximately 6 times every 100 years the race will be run with two undefeated horses.

POI 9: RELIABILITY OF STANDBY REDUNDANT SYSTEM

Consider a standby redundancy system with one operating unit and one on standby, i.e., a system that can survive one failure. If the failure rate is 2 units per year, what is the 6-month reliability of the system?

Solution

As discussed earlier, a standby redundant system is one in which one unit is in the operating mode and n identical units are in standby mode. Unlike a parallel system in which all units in the system are active, in the standby redundancy system the standby units are inactive. If all units have the same failure rate in the operating mode, unit failures are independent of each other, standby units have zero failure rate in the standby mode, and there is perfect switchover to a standby unit when the operating unit fails, then the reliability R of the standby redundancy system is given by

$$R = \sum_{x=0}^{n} \frac{e^{-\mu}(\mu)^x}{x!} \tag{15.3}$$

This represents the probability of n or fewer failures in a time period on which the average number of failures is μ.

The required reliability is a 6-month reliability. Therefore, the associated time period is 6 months. The given failure rate is 2 units per year. Therefore, the associated time period is 1 year or 12 months. Note that 6 divided by 12 yields 1/2. Multiply this result by the failure rate to obtain μ:

$$\mu = (1/2)(2) = 1$$

Substitute the value of μ in the Poisson pdf:

$$f(x) = \frac{e^{-1}(1)^x}{x!}; \quad x = 0, 1, 2, \ldots$$

Identify n, the number of units in standby mode, $n = 1$.

Compute the required reliability by summing the Poisson pdf from 0 to n. The 6-month reliability is given by Equation (15.3)

$$R = \sum_{x=0}^{1} \frac{e^{-1}(1)^x}{x!}$$

$$= 0.368 + 0.368$$

$$= 0.736$$

POI 10: Bacteria Growth

Bacteria are known to be present in a source of liquid with a mean number of 3 per cubic centimeter. Ten 1-cm^3 test tubes are filled with the liquid. Calculate the probability that all ten test tubes will show growth (i.e., contain at least one bacterium each).

Solution

Let X denote the number of bacteria in 1 cm^3. Substitute $\mu = 3$ in the Poisson pdf gives:

$$f(x) = \frac{e^{-3}(3)^x}{x!}; \quad x = 0, 1, 2, \ldots$$

The probability that a test tube will show growth is

$$P(X \geq 1) = 1 - P(X = 0)$$

$$= 1 - e^{-3}$$

$$= 1 - 0.0498$$

$$= 0.9502$$

The probability that all ten test tubes will show growth is

$$P(X = 10) = (0.9502)^{10}$$

$$= 0.60$$

POI 11: Radioactive Emission Rate

Assume the number of particles emitted by a radioactive substance has a Poisson distribution with an average emission of one particle per second.

1. Find the probability that at most one particle will be emitted in 3 s.
2. How low an emission rate would be required to make the probability of the emission of at most one particle in 3 s at least 0.80?

Solution

1. Let X denote the number of radioactive particles emitted in 3 s. Because $\lambda = 1$ and $t = 3$, $\mu = 3$. Therefore, the pdf of X is

$$f(x) = \frac{e^{-3}(3)^x}{x!}; \quad x = 0, 1, 2, \dots$$

The probability that at most one particle will be emitted in 3 s is

$$P(X \le 1) = \sum_{x=0}^{1} \frac{e^{-3}(3)^x}{x!}$$

$$= P(X = 0) + P(X = 1)$$

$$= e^{-3} + e^{-3}(3)$$

$$= 0.0498 + 0.1494$$

$$= 0.1992$$

2. Let λ be the emission rate per second. Then the pdf of X is

$$f(x) = \frac{e^{-3\lambda}(3\lambda)^x}{x!}$$

In addition,

$$P(X \le 1) = \sum_{x=0}^{1} \frac{e^{-3\lambda}(3\lambda)^x}{x!}$$

$$= P(X = 0) + P(X - 1)$$

$$= e^{-3\lambda}(1 + 3\lambda)$$

For the probability of emission of at most one particle in 3 s to be at least 0.80, λ, the emission rate per second must be such that

$$e^{-3\lambda}(1 + 3\lambda) \geq 0.80$$

This will be solved using a trial-and-error method.
For $\lambda = 0.3$,

$$e^{-3\lambda}(1 + 3\lambda) - 0.80 = -0.0275$$

For $\lambda = 0.2$,

$$e^{-3\lambda}(1 + 3\lambda) - 0.80 = 0.0781$$

Using Newton's method or the equivalent, 0.275 is obtained as an improved approximation to the root of the equation,

$$e^{-3\lambda}(1 + 3\lambda) - 0.80 = 0$$

Therefore, an emission rate per second of 0.275 or less is required to make the probability of the emission of at most one particle in 3 s at least 0.80.

POI 12: Hazardous Waste Incinerator Truck Scheduling

The number of hazardous waste trucks arriving daily at a certain hazardous waste incineration facility has a Poisson distribution with parameter $n = 2.5$. Present facilities can accommodate three trucks a day. If more than three trucks arrive in a day, the trucks in excess of three must be sent elsewhere.

1. On a given day, what is the probability of having to send a truck elsewhere?
2. How much must the present waste facilities be increased to permit handling of all trucks on about 95% of the days?

Solution

1. Let X denote the number of trucks on a given day. Then

$$P(\text{sending trucks elsewhere}) = P(X \geq 4)$$

This may also be written as

$$P(X \geq 4) = 1 - P(X < 4) = 1 - P(X \leq 3)$$

$$P(X \geq 4) = 1 - P(X = 0) - P(X = 1) - P(X = 2) - P(X = 3)$$

Employing the Poisson pdf with $\mu = 2.5$ gives

$$P(X \geq 4) = 1 - \left(e^{-2.5} + 2.5e^{-2.5} + \frac{(2.5)^2 e^{-2.5}}{2} + \frac{(2.5)^3 e^{-2.5}}{6} \right)$$

$$= 1 - (0.0821 + 0.2052 + 0.2565 + 0.2138)$$

$$= 0.2424$$

2. Note that the number of trucks has not been specified. The general solution is given by

$$P(\text{sending trucks elsewhere}) = 0.05 = 1 - \sum_{i=0}^{n} P(X = i)$$

A trial-and-error solution is again required because a different equation is obtained for each value of n. For example, if $n = 5$,

$$0.05 = 1 - [P(X = 0) + P(X = 1) + P(X = 2) + P(X = 3) + P(X = 4) + P(X = 5)]$$

Substituting gives

$$0.05 = 1 - e^{-2.5}\left[1 + 2.5 + \frac{(2.5)^2}{2} + \frac{(2.5)^3}{6} + \frac{(2.5)^4}{24} + \frac{(2.5)^5}{120}\right]$$

$$= 1 - 0.0821[1.0 + 2.5 + 3.125 + 2.604 + 1.628 + 0.814]$$

$$= 1 - 0.958$$

$$0.5 \approx 0.042$$

The increase is from 3 to 5; therefore, the facilities must be increased by approximately 67%.

16 EXP. Exponential Distribution

INTRODUCTION

The exponential distribution is an important distribution in that it represents the distribution of the time required for a single event from a Poisson process to occur. In particular, in sampling from a Poisson distribution with parameter μ, the probability that no event occurs during $(0,t)$ is $e^{-\lambda t}$. Consequently, the probability that an event will occur during $(0,t)$ is

$$F(t) = 1 - e^{-\lambda t} \tag{16.1}$$

This represents the cumulative distribution function (cdf) of t. One can therefore show that the probability distribution function (pdf) is

$$f(t) = e^{-\lambda t} \tag{16.2}$$

Note that the parameter $1/\lambda$ (sometimes denoted as μ) is the expected value. Normally, the reciprocal of this value is specified and represents the expected value of $f(t)$.

Because the exponential function appears in the expression for both the pdf and cdf, the distribution is justifiably called the *exponential distribution*. A typical pdf of x plot is provided in Figure 16.1.

Alternatively, the cumulative exponential distribution can be obtained from the pdf (with x replacing t):

$$F(x) = \int_0^x \lambda e^{-\lambda x} \, dx = 1 - e^{-\lambda x} \tag{16.3}$$

All that remains is a simple evaluation of the negative exponent in Equation 16.3.

In statistical and reliability applications one often encounters a random variable's conditional failure density or hazard function, $g(x)$. In particular, $g(x) \, dx$ is the probability that a "product" will fail during $(x, x + dx)$ under the condition that it had not failed before time x. Consequently,

$$g(x) = \frac{f(x)}{1 - F(x)} \tag{16.4}$$

If the probability density function $f(x)$ is exponential, with parameter λ, it follows from Equation 16.2 and Equation 16.3 that

$$g(x) = \frac{\lambda e^{-\lambda x}}{1 - (1 - e^{-\lambda x})}$$

$$= \frac{\lambda e^{-\lambda x}}{e^{-\lambda x}} \tag{16.5}$$

$$g(x) = \lambda$$

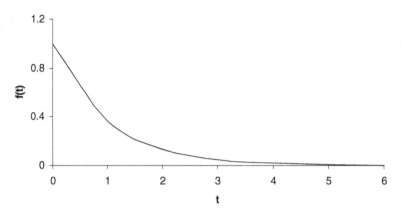

FIGURE 16.1 Exponential distribution.

Equation 16.5 indicates that the failure probability is constant, irrespective of time. It implies that the probability that a component whose time-to-failure distribution is exponential fails in an instant during the first hour of its life is the same as its failure probability during an instant in the thousandth hour — presuming it has survived up to that instant. It is for this reason that the parameter λ is usually referred to in life-test applications as the *failure rate*. This definition generally has meaning only with an exponential distribution. Other failure rate definitions and applications will appear later in Section II, particularly in Chapter 17.

This natural association with life-testing and the fact that it is very tractable mathematically makes the exponential distribution attractive as representing the life distribution of a complex system or several complex systems. In fact, the exponential distribution is as prominent in reliability analysis as the normal distribution is in other branches of statistics.

PROBLEMS AND SOLUTIONS

EXP 1: EXPONENTIAL DISTRIBUTION ADVANTAGES AND LIMITATIONS

Discuss some of the advantages and limitations of the exponential distribution.

Solution

It has been shown theoretically that this distribution provides a reasonable model for systems designed with a limited degree of redundancy and made up of many components, none of which has a high probability of failure. This is especially true when low component failure rates are maintained by periodic inspection and replacement or in situations in which failure is a function of outside phenomena rather than a function of previous conditions. On the other hand, the exponential distribution often cannot represent individual component life (because of "infant mortalities" and wear out patterns — see Chapter 17), and it is sometimes questionable even as a system model.

EXP 2: PUMP EXPECTED LIFE

Estimate the probability that a pump will survive at least three times its expected life. Assume the exponential distribution will apply.

Solution

The exponential distribution gives

$$P(T) = e^{-\lambda t}$$

with $\lambda = 1/a$ and $t = 3a$, where a = expected life of the pump. Thus,

$$P(T > 3a) = e^{-\left(\frac{1}{a}\right)(3a)}$$

$$= e^{-3}$$

$$= 0.0498 = 4.98\%$$

Therefore, there is a 5% chance that the pump will survive past three times its expected life.

EXP 3: PUMP EXPECTED LIFE INCREASE

Refer to Problem EXP 2. Calculate the probability that a pump will survive:

1. At least five times its expected life
2. At least ten times its expected life

Solution

The describing equation

$$P(T) = e^{-\lambda t}$$

remains the same.
    For case 1,

$$t = 5a$$

so that

$$P(T > 5a) = e^{-\left(\frac{1}{a}\right)(5a)}$$

$$= e^{-5}$$

$$= 0.0067 = 0.67\%$$

Similarly, for case 2,

$$P(T > 10a) = e^{-\left(\frac{1}{a}\right)(10a)}$$

$$= e^{-10}$$

$$= 4.54 \times 10^{-5} = 4.54 \times 10^{-3}\%$$

As expected, the probability decreases with increasing survival time.

EXP 4: EXPECTED LIFE

The time to failure for a battery is presumed to follow an exponential distribution with $\lambda = 0.1$ (per year). What is the probability of a failure within the first year?

Solution

Refer to Equation 16.3 in the introduction to this chapter.

$$F(x) = \int_0^x \lambda e^{-\lambda x}\, dx = 1 - e^{-\lambda x}$$

For this case,

$$P(X \le 1) = \int_0^1 (0.1)e^{-(0.1)x}\, dx$$

$$= -\frac{0.1}{0.1} e^{-(0.1)x}\Big|_0^1$$

$$= -e^{-0.1} + e^0$$

$$= 1 - e^{-0.1}$$

$$= 0.095 = 9.5\%$$

Therefore, there is nearly a 10% probability that the battery will fail within the first year.

EXP 5: ELECTRONIC SYSTEM IN SERIES

An electronic system consists of three components (1, 2, 3) connected in parallel. If the time to failure for each component is exponentially distributed and mean times of failures for components 1, 2, and 3, are 200, 300, and 600 d, respectively, determine the system reliability for 365 d.

Solution

For this series system, the probability of failure is (see Chapter 11)

$$P(F) = P(\text{all components fail})$$

$$= (1 - P_1)(1 - P_2)(1 - P_3)$$

where P_i is the probability of surviving 365 d. The system reliability is

$$R = 1 - P(F)$$

$$= 1 - (1 - P_1)(1 - P_2)(1 - P_3)$$

Based on the data provided

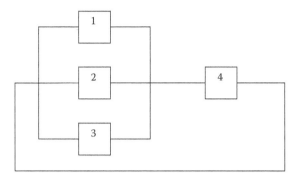

FIGURE 16.2 Pumping system.

$$P_1 = e^{-\left(\frac{365}{200}\right)} = 0.161$$

$$P_2 = e^{-\left(\frac{365}{300}\right)} = 0.296$$

$$P_3 = e^{-\left(\frac{365}{600}\right)} = 0.544$$

Therefore, the system reliability is

$$R = 1 - (1 - 0.161)(1 - 0.296)(1 - 0.544)$$

$$= 0.731 = 73.1\%$$

EXP 6: FOUR-COMPONENT PUMPING SYSTEM

A pumping system consists of four components. Three are connected in parallel, which in turn are connected downstream (in series) with the other component. The arrangement is schematically shown in Figure 16.2.

If the pumps have the same exponential failure rate, λ, of 0.5 (year)$^{-1}$, estimate the probability that the system will not survive for more than one year.

Solution

Based on the information provided, the pumping system fails when the three parallel components fail or when the downstream component fails. This is a combination of a parallel and series system. From Equation 13.2 in Chapter 13, the reliability is

$$R(t) = 1 - F(t)$$

where $F(t)$ is the probability of failure between 0 and t. From Equation 11.2 in Chapter 11, the reliability of the parallel system is

$$R_p = 1 - (1 - R_1)(1 - R_2)(1 - R_3); \ R_i = R$$

$$= 1 - (1 - R)^3$$

From Equation 11.1 in Chapter 11, for a series system

$$R_s = R_p R_4 = R_p R$$

where R_s also represents the overall sytem reliability.

Applying the exponential model gives

$$R(t) = 1 - F(t)$$

$$= 1 - (1 - e^{-\lambda t})$$

$$= e^{-\lambda t}$$

$$R_1 = R_2 = R_3 = R_4 = R = e^{-(0.5)(1)}$$

$$R = 0.6065$$

Thus,

$$R_p = 1 - (1 - 0.6065)^3$$

$$= 1 - 0.0609$$

$$= 0.9391$$

and

$$R_s = (0.9391)(0.6065)$$

$$= 0.57 = 57\%$$

EXP 7: REPLACEMENT TIME I

The probability that a thermometer in a hazardous waste incinerator will not survive for more than 36 months is 0.925; how often should the thermometer be replaced? Assume the time to failure is exponentially distributed and that the replacement time should be based on the thermometer's expected life.

Solution

This requires the calculation of μ in the exponential model with units of (month)$^{-1}$. Once again,

$$F(t) = 1 - e^{-\lambda t}$$

Based on the information provided

$$P(T \leq 36) = 0.925$$

or

$$0.925 = 1 - e^{-(\lambda)(36)}$$

Solving for gives

$$\lambda = -\frac{1}{36}\ln(1-0.925)$$

$$= 0.07195$$

Because the expected time (or life), $E(T)$, is

$$E(T) = \frac{1}{\lambda}$$

$$= \frac{1}{0.07195}$$

$$= 13.9 \text{ months}$$

Therefore, the thermometer should be replaced in approximately 14 months.

EXP 8: REPLACEMENT TIME II

Refer to Problem EXP 7. Determine when the thermometer should be replaced if the probability of thermometer survival (for 36 months) is

1. 0.95
2. 0.99

Solution

The describing equation remains

$$F(t) = 1 - e^{-\lambda t}$$

1. For $F(t) = 0.95$,

$$0.95 = 1 - e^{-(\lambda)(36)}$$

Solving for μ gives

$$\lambda = -\frac{1}{36}\ln(1-0.95)$$

$$= 0.0832$$

The expected life is

$$E(T) = \frac{1}{\lambda}$$

$$= \frac{1}{0.0832}$$

$$= 12 \text{ months}$$

2. Similarly,

$$\lambda = -\frac{1}{36}\ln(1-0.99)$$

$$= 0.128$$

and

$$E(T) = \frac{1}{\lambda}$$

$$= \frac{1}{0.128}$$

$$= 7.82 \text{ months}$$

17 WBL. Weibull Distribution

INTRODUCTION

Unlike the exponential distribution, the failure rate of equipment frequently exhibits three stages: a break-in stage with a declining failure rate, a useful life stage characterized by a fairly constant failure rate, and a wear out period characterized by an increasing failure rate. Many industrial parts and components follow this path. A failure rate curve exhibiting these three phases (see Figure 17.1) is called a *bathtub* curve.

In the case of the bathtub curve, failure rate during useful life is constant. Letting this constant be α and substituting it for $Z(t)$ in Equation 13.18 yields

$$F(t) = \alpha \exp\left(-\int_0^t \alpha\, dt\right)$$

$$= \alpha \exp(-\alpha t); \quad t > 0 \qquad (17.1)$$

$$= \alpha e^{-\alpha t}$$

as the probability distribution function (pdf) of time to failure during the useful life stage of the bathtub curve. Equation 17.1 defines an exponential pdf (see Equation 16.2) that is, a special case of the pdf defining the Weibull distribution.

Weibull introduced the distribution, which bears his name principally on empirical grounds, to represent certain life-test data. The Weibull distribution provides a mathematical model of all three stages of the bathtub curve. This is now discussed. An assumption about the failure rate that reflects all three stages of the bathtub stage is

$$Z(t) = \alpha\beta t^{\beta-1}; \quad t > 0 \qquad (17.2)$$

where α and β are constants. For $\beta < 1$ the failure rate $Z(t)$ decreases with time. For $\beta = 1$ the failure rate is constant and equal to α. For $\beta > 1$ the failure rate increases with time. Using Equation 13.18 again to translate the assumption about failure rate into a corresponding assumption about the pdf of T, time to failure, one obtains

$$f(t) = \alpha\beta t^{\beta-1} \exp\left(\int_0^t \alpha\beta t^{\beta-1}\, dt\right)$$

$$= \alpha\beta t^{\beta-1} \exp(-\alpha t^\beta); \quad t > 0; \quad \alpha > 0, \beta > 0 \qquad (17.3)$$

Equation 17.3 defines the pdf of the Weibull distribution. The exponential distribution discussed in the preceding text, whose pdf is given in Equation 16.2 earlier, is a special case of the Weibull distribution with $\beta = 1$. The variety of assumptions about failure rate and the probability distribution

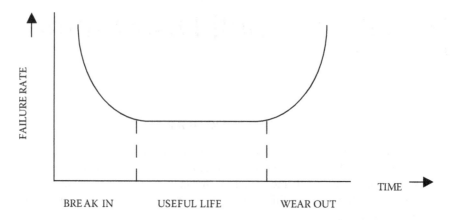

FIGURE 17.1 Bathtub curve.

of time to failure that can be accommodated by the Weibull distribution make it especially attractive in describing failure time distributions in industrial and process plant applications.

To illustrate probability calculations involving the exponential and Weibull distributions introduced in conjunction with the bathtub curve of failure rate, consider first the case of a transistor having a constant rate of failure of 0.01 per thousand hours. To find the probability that the transistor will operate for at least 25,000 h, substitute the failure rate

$$Z(t) = 0.01$$

into Equation 13.16 in Chapter 13, which yields

$$f(t) = \exp\left(-\int_0^t 0.01\,dt\right)$$

$$= 0.01e^{-0.01t}; \quad t > 0$$

as the pdf of T, the time to failure of the transistor. Because t is measured in thousands of hours, the probability that the transistor will operate for at least 25,000 h is given by

$$P(T > 25) = \int_{25}^{\infty} -0.01e^{-0.01t}\,dt$$

$$= -e^{-\infty} + e^{-0.01(25)}$$

$$= 0 + 0.78$$

$$= 0.78$$

Now suppose it is desired to determine the 10,000-h reliability of a circuit of five such transistors connected in series. The 10,000-h reliability of one transistor is the probability that it will last at least 10,000 h. This probability can be obtained by integrating the pdf of T, time to failure, which gives

$$P(T > 10) = \int_{10}^{\infty} -0.01 e^{-0.01t} \, dt$$

$$= -e^{-\infty} + e^{-0.01(10)}$$

$$= 0 + 0.90$$

$$= 0.90$$

The same result can also be obtained directly from Equation 13.15 in Chapter 13, which expresses reliability in terms of failure rate. Substituting the failure rate

$$Z(t) = 0.01$$

into this equation yields

$$R(t) = \exp\left(-\int_{0}^{t} 0.01 \, dt \right)$$

as the reliability function. The 10,000-h reliability is therefore

$$R(10) = \exp\left(-\int_{0}^{10} 0.01 \, dt \right)$$

$$= e^{-0.01(10)}$$

$$= e^{-0.1}$$

$$= 0.90$$

The 10,000-h reliability of a circuit of 5 transistors connected in series is obtained by applying the formula for the reliability of series system (see Section SPS), to obtain

$$R_s = [R(10)]^5$$

$$= (0.9)^5$$

$$= 0.59$$

As another example of probability calculations, consider a component whose time to failure T, in hours, has a Weibull pdf with parameters $\alpha = 0.01$ and $\beta = 0.50$ in Equation 17.3. This gives

$$f(t) = (0.01)(0.50)t^{0.5-1} e^{-(0.01)t^{0.5}}; \quad t > 0$$

as the Weibull pdf of the failure time of the component under consideration. The probability that the component will operate at least 8100 h is then given by

$$P(T > 8100) = \int_{8100}^{\infty} f(t)\,dt$$

$$= \int_{8100}^{\infty} 0.005t^{-0.5} e^{-(0.01)t^{0.5}}\,dt$$

$$= e^{-0.01t^{0.5}} \Big|_{8100}^{\infty}$$

$$= 0 + e^{-(0.01)(8100)^{0.5}}$$

$$= 0.41$$

PROBLEMS AND SOLUTIONS

WBL 1: Weibull Distribution Applications

Discuss some Weibull distribution applications.

Solution

A variety of conditional failure distributions, including wearout patterns, can be accommodated by the Weibull distribution. Therefore, this distribution has been frequently recommended — instead of the exponential distribution — as an appropriate failure distribution model. Empirically, satisfactory fits have been obtained to failure data on electron tubes, relays, ball bearings, metal fatigue, and even business mortality.

WBL 2: Constant Rate Equation

"Derive" Equation 17.1.

Solution

Equation 13.18 earlier was shown to take the form

$$f(t) = Z(t)\exp\left[-\int_0^t Z(t)\,dt\right]$$

During the constant rate $Z(t)$ is given by

$$Z(t) = \alpha = \text{constant}$$

Substituting into Equation 13.18 gives

$$f(t) = \alpha\exp\left[-\int_0^t \alpha\,dt\right]$$

$$= \alpha\exp(-\alpha t)$$

$$= \alpha e^{-\alpha t}$$

WBL 3: Failure Probability from Failure Rate

The life (time to failure) of a machine component has a Weibull distribution. Outline how to determine the probability that the component lasts a given period of time if the failure rate is $t^{-1/2}$.

Solution

Identify the failure rate, $Z(t)$, from Equation 17.2.

$$Z(t) = t^{-1/2}$$

Also, identify the values of and appearing in the failure rate. If the failure rate is $t^{-1/2}$,

$$\beta - 1 = -1/2$$

and

$$\alpha\beta = 1$$

Therefore, $\beta = 1/2$ and $\alpha = 2$.
For these values of α and β obtained above, determine the Weibull pdf:

$$f(t) = t^{-1/2}e^{-2t^{1/2}}; \quad t > 0$$

Integration of this pdf will yield the required probability. This is demonstrated in Problem WBL 4.

WBL 4: Failure Time Calculation

Refer to Problem WBL 3. Determine the probability that the component lasts at least 25,000 h if t is measured in thousands of hours.

Solution

For this case,

$$t = 25$$

Because time is measured in thousands of hours the probability that the component lasts at least 25,000 h is

$$P(T > 25) = \int_{25}^{\infty} t^{-1/2} e^{-2t^{1/2}} \, dt$$

$$= -e^{-2\sqrt{t}} \rbrack_{25}^{\infty}$$

This may be integrated to give

$$P(T > 25) = -0 - (-e^{-2(25)^{0.5}})$$

$$= 4.5 \times 10^{-5}$$

WBL 5: Different Weibull Coefficients

Refer to Problem WBL 3. Obtain the probability equation if the failure rate is given by

1. $t^{-0.25}$
2. $t^{-\sqrt{1/2}}$

Solution

For case 1,

$$\beta - 1 = -0.25$$

$$\beta = 0.75$$

Because

$$\alpha\beta = 1$$

$$\alpha(0.75) = 1$$

$$\alpha = 1.33$$

2. For case 2,

$$\beta - 1 = -\sqrt{1/2}$$

$$\beta = 0.293$$

and

$$\alpha\beta = 1$$

$$\alpha(0.293) = 1$$

$$\alpha = 3.41$$

WBL 6: Gasket Life

The life of a gasket has a Weibull distribution with failure rate

$$Z(t) = \frac{1}{\sqrt{t}}$$

where t is measured in years.

What is the probability that the gasket will last at least 4 years?

Solution

The pdf specified by $f(t)$ in terms of the failure rate, $Z(t)$, is as follows (see Introduction):

$$f(t) = Z(t) \exp\left(-\int_0^t Z(t)\, dt\right)$$

Substituting $1/t^{1/2}$ for $Z(t)$ yields

$$f(t) = t^{-1/2} \exp\left(-\int_0^\infty t^{-1/2}\, dt\right)$$

$$= t^{-1/2} \exp(-2t^{1/2})\, dt; \quad t > 0$$

Employ the integration procedure in the Introduction and in Problem WBL 4. The probability that the seal lasts at least 4 years is

$$P(T \geq 4) = \int_4^\infty t^{-1/2} \exp(-2t^{1/2})\, dt$$

$$= [-e^{-2t^{0.5}}]_4^\infty$$

$$= -0 - (-e^{-4})$$

$$= 0.0183$$

WBL 7: ELECTRONIC COMPONENT AVERAGE LIFE

The life of an electronic component is a random variable having a Weibull distribution with $\alpha = 0.025$ and $\beta = 0.50$. What is the average life of the component?

Solution

Let T denote the life in hours of the electronic component. The pdf of T is again obtained by applying Equation 17.3, which yields

$$f(t) = \alpha\beta t^{\beta-1} \exp(-\alpha t^\beta); \quad t > 0; \; \alpha > 0, \; \beta > 0$$

Substituting $\alpha = 0.025$ and $\beta = 0.50$ yields

$$f(t) = (0.025)(0.50)t^{0.50-1} \exp(-0.025 t^{0.5}); \quad t > 0$$

$$= (0.0125)t^{0.50-1} \exp(-0.025 t^{0.5}); \quad t > 0$$

The average value of T is given by integration of Equation 8.1:

$$E(T) = \int_{-\infty}^\infty t\, f(t)\, dt$$

$$= \int_0^\infty 0.0125 t^{0.50} \exp(-0.025 t^{0.5})\, dt$$

Integrating as before,

$$E(T) = 3200 \text{ h}$$

WBL 8: ELECTRONIC COMPONENT LIFE

Refer to Problem WBL 7. What is the probability that the component will last more than 4000 h?

Solution

The probability that the component will last more than 4000 h is given by

$$P(T > 4000) = \int_{4000}^{\infty} f(t)\,dt$$

$$= \int_{4000}^{\infty} (0.0125)t^{0.50-1} \exp(-0.025t^{0.5})\,dt$$

$$= [-e^{-0.025t^{0.5}}]_{4000}^{\infty}$$

$$= -0 - (-e^{-1.581})$$

$$= 0.2057$$

WBL 9: OUTLINE FOR ESTIMATING WEIBULL PARAMETERS

Develop an outline for estimating the Weibull parameters.

Solution

Estimation of the parameters in the pdf of the Weibull distribution,

$$f(t) = \alpha\beta t^{\beta-1} \exp(-\alpha t^{\beta}); \ t > 0; \ \alpha > 0, \ \beta > 0$$

can be obtained by a graphical procedure given by Bury [1]. It is based on the fact that

$$\ln\left[\ln\frac{1}{1-F(t)}\right] = \ln\alpha + \beta\ln t \tag{17.4}$$

is a linear function of ln t. Here,

$$F(t) = 1 - e^{-\alpha t^{\beta}}; \quad t > 0 \tag{17.5}$$

$$= 0; \quad t < 0$$

defines the cdf of the Weibull distribution. The graphical procedure for estimating the slope, β, and the intercept, ln α, of time to failure, involves first ordering the observations from smallest ($i = 1$) to largest ($i = n$). The value of the ith observation varies from sample to sample. It can be shown that the average value of $F(t)$ for t equal to the value of the ith observation on T is $i/(n+1)$. The points obtained by plotting

---

TABLE 17.1
Weibull Coefficients

(1) Time to Failure (*t*)	(2) Order of Failure (*i*)	(3) ln *t*	(4) ln{ln 1/[1 − *i*/(*n* + 1)]}
18	1	2.89	2.35
36	2	3.58	1.61
40	3	3.69	1.14
53	4	3.97	0.79
71	5	4.26	0.50
90	6	4.50	0.24
106	7	4.66	0.01
127	8	4.84	0.26
149	9	5.00	0.53
165	10	5.11	0.87

$$\ln \left[\ln \frac{1}{1 - \dfrac{i}{n+1}} \right] \qquad (17.6)$$

against the natural logarithm of the *i*th observation for *i* = 1 to *i* = *n* should lie along a straight line whose slope is β and whose intercept is ln α if the assumption that *T* has a Weibull distribution is correct. This procedure is demonstrated in Problem WDL 10.

WBL 10: ESTIMATION OF WEIBULL PARAMETERS

The time in days to failure of each sample of ten electronic components is observed as follows:

$$71, 40, 90, 149, 127, 53, 106, 36, 18, 165$$

Assuming a Weibull distribution applies, estimate the parameters α and β.

Solution

The observations in order of magnitude are

$$18, 36, 40, 53, 71, 90, 106, 127, 149, 165$$

Table 17.1 is generated from the data. Set $X = \ln t$ and $Y = \ln\{\ln 1/[1 - i/(n + 1)]\}$; values of X are in column (3) and values of Y are in column (4).

Compute the values of $\sum_{i=1}^{n} X$. Similarly, calculate $\sum X^2$, $\sum Y$, $\sum XY$. These values determine the slope and intercept of the least-squares line of best fit to estimate Y from X. Details of this procedure are provided in Chapter 28 (Section III):

$$\sum X = 42.5 \quad \sum Y = -4.96 \quad n = 10$$

$$\sum X^2 = 185.2 \quad \sum XY = -14.71$$

For example,

$$\sum_{i=1}^{10} (X)^2 = (2.89)^2 + (3.58)^2 + (3.69)^2 + (3.97)^2 + (4.26)^2 + (4.50)^2 + (4.66)^2 \\ + (4.84)^2 + (5.00)^2 \qquad + (5.11)^2$$

$$= 185.2$$

Substitute these values in the following formulas for the slope and intercept of the least-squares line of best fit:

$$\text{Slope} = \frac{n \sum XY - \left(\sum X\right)\left(\sum Y\right)}{n \sum X^2 - \left(\sum X\right)^2}$$

$$= \frac{(10)(-14.71) - (42.5)(-4.96)}{(10)(185.2) - (42.5)^2}$$

$$= 1.4$$

$$\text{Intercept} = \frac{\sum Y}{n} - (slope)\frac{\sum X}{n}$$

$$= \frac{-4.96}{10} - (1.4)\frac{42.5}{10}$$

$$= -6.4$$

The estimated value of β is 1.4, and the estimated value of α is the antilog of -6.4, which is equal to 0.0017.

REFERENCE

1. Bury, K., *Statistical Models in Applied Science*, John Wiley & Sons, Hoboken, NJ, 1975.

18 NOR. Normal Distribution

INTRODUCTION

The initial presentation in this chapter on normal distributions will focus on failure rate, but can be simply applied to all other applications involving normal distributions. When T, time to failure, has a normal distribution, its probability distribution function (pdf) is given by

$$f(t) = \frac{1}{\sqrt{2\pi}\,\sigma} \exp\left[-\frac{1}{2}\left(\frac{t-\mu}{\sigma}\right)^2\right]; \quad -\infty < t < \infty \tag{18.1}$$

where μ is the mean value of T and σ is its standard deviation. The graph of $f(t)$ is the familiar bell-shaped curve shown in Figure 18.1.

The reliability function corresponding to the normally distributed failure time is given by

$$R(t) = \frac{1}{\sqrt{2\pi}\,\sigma} \int_t^\infty \exp\left[-\frac{1}{2}\left(\frac{t-\mu}{\sigma}\right)^2\right] dt \tag{18.2}$$

The corresponding failure rate (hazard rate) is obtained by substitution of Equation 18.2 into Equation 13.11 and Equation 13.12 in Chapter 13, which state that the failure rate $Z(t)$ is related to the reliability $R(t)$ by

$$Z(t) = \frac{R'(t)}{R(t)}$$

Recalling that Equation 13.11 in Chapter 13 states

$$R'(t) = F'(t)$$

And substituting f(t) from Equation 18.1 and R(t) from Equation 18.2 yields

$$Z(t) = \frac{\exp\left[-\frac{1}{2}\left(\frac{t-\mu}{\sigma}\right)^2\right]}{\int_t^\infty \exp\left[-\frac{1}{2}\left(\frac{t-\mu}{\sigma}\right)^2\right] dt} \tag{18.3}$$

as the failure rate corresponding to a normally distributed failure time.

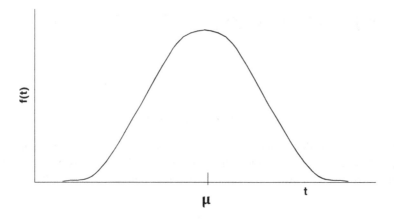

FIGURE 18.1 Normal pdf of time to failure.

If T is normally distributed with mean μ and standard deviation σ, then the random variable $(T - \mu)/\sigma$ is normally distributed with mean 0 and standard deviation 1. The term $(T - \mu)/\sigma$ is called a *standard normal curve* that is represented by Z, not to be confused with the failure rate $Z(t)$. This is discussed in more detail later in this chapter.

Table A.1A (Appendix A) is a tabulation of areas under a standard normal curve to the right of z_0 for nonnegative values of z_0. Probabilities about a standard normal variable Z can be determined from this table. For example,

$$P(Z > 1.54) = 0.062$$

is obtained directly from Table A.1A (Appendix A) as the area to the right of 1.54. As presented in Figure 18.2, the symmetry of the standard normal curve about zero implies that the area to the right of zero is 0.5, and the area to the left of zero is 0.5.

Plots demonstrating the effect of μ and σ on the bell-shaped curve are provided in Figure 18.3 and Figure 18.4.

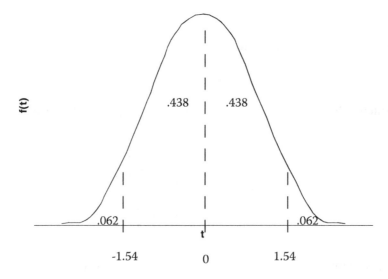

FIGURE 18.2 Areas under a standard normal curve.

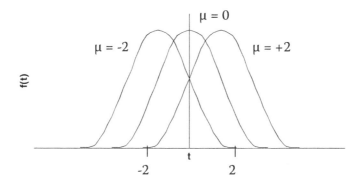

FIGURE 18.3 Normal pdf — varying μ.

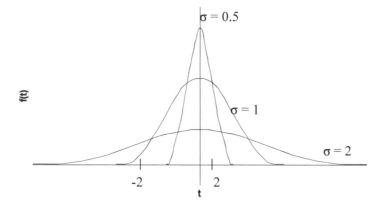

FIGURE 18.4 Normal pdf — varying σ.

Consequently, one can deduce from Table A.1A of Appendix A and Figure 18.2 that

$$P(0 < Z < 1.54) = 0.5 - 0.062 = 0.438$$

Also, because of symmetry

$$P(-1.54 < Z < 0) = 0.438$$

and

$$P(Z < -1.54) = 0.062$$

The following probabilities can also be deduced by noting that the area to the right of 1.54 is 0.062 in Figure 18.2.

$$P(-1.54 < Z < 1.54) = 0.876$$

$$P(Z < 1.54) = 0.938$$

$$P(Z > -1.54) = 0.938$$

Table A.1A (Appendix A) can also be used to determine probabilities concerning normal random variables that are not standard normal variables. The required probability is first converted to an equivalent probability about a standard normal variable. For example, if T, the time to failure, is normally distributed with mean $\mu = 100$ and standard deviation $\sigma = 2$ then $(T - 100)/2$ is a standard normal variable and one may write

$$P(T_1 < T < T_2) = P\left(\frac{T_1 - \mu}{\sigma} < \frac{T - \mu}{\sigma} < \frac{T_2 - \mu}{\sigma}\right)$$

where

$$\frac{T - \mu}{\sigma} = Z = \text{ standard normal variable} \qquad (18.4)$$

Therefore, if $T_1 = 98$ and $T_2 = 104$ for this example, the describing equation becomes

$$P(98 < T < 104) = \left(\frac{98 - \mu}{\sigma} < \frac{T - \mu}{\sigma} < \frac{104 - \mu}{\sigma}\right)$$

$$= P\left(\frac{98 - 100}{2} < \frac{T - \mu}{2} < \frac{104 - 100}{2}\right)$$

$$= P\left(-1 < \frac{T - 100}{2} < 2\right)$$

$$= P(-1 < Z < 2)$$

$$= 0.341 + 0.477$$

$$= 0.818$$

For any random variable X — where X has replaced T — that is normally distributed with mean μ and standard deviation, one may now write as follows:

$$P(\mu - \sigma < X < \mu + \sigma) = P\left(-1 < \frac{X - \mu}{\sigma} < 1\right)$$
$$= P(-1 < Z < 1) = 0.68 \qquad (18.5)$$

$$P(\mu - 2\sigma < X < \mu + 2\sigma) = P\left(-2 < \frac{X - \mu}{\sigma} < 2\right)$$
$$= P(-2 < Z < 2) = 0.95 \qquad (18.6)$$

$$P(\mu - 3\sigma < X < \mu + 3\sigma) = P\left(-3 < \frac{X - \mu}{\sigma} < 3\right)$$
$$= P(-3 < Z < 3) = 0.997 \qquad (18.7)$$

The probabilities given in Equation 18.5 to Equation 18.7 are the sources of the percentages cited earlier. These can be used to interpret the standard deviation s of a sample of observations on a normal random variable, as a measure of dispersion about the sample mean $\overline{X}$.

The normal distribution is used to obtain probabilities concerning the mean $\overline{X}$ of a sample of n observations on a random variable X. If X is normally distributed with mean μ and standard deviation σ, then $\overline{X}$, the sample mean, is normally distributed with mean μ and standard deviation $\sigma/\sqrt{n}$. For example, suppose X is normally distributed with mean 100 and standard deviation 2. Then $\overline{X}$, the mean of a sample of 16 observations on X, is normally distributed with mean 100 and standard deviation 0.5. To calculate the probability that $\overline{X}$ is greater than 101, one would write

$$P(\overline{X} > 101) = P\left[\frac{\overline{X} - 100}{0.5} > \frac{101 - 100}{0.5}\right]$$

$$P(\overline{X} > 101) = P\left[\frac{\overline{X} - 100}{0.5} > \frac{101 - 100}{0.5}\right]; \quad Z = \frac{\overline{X} - 100}{0.5}$$

$$= P(Z > 2)$$

$$= 0.023$$

If X is not normally distributed, then $\overline{X}$, the mean of a sample of n observations on X, is approximately normally distributed with mean μ and standard deviation $\sigma/\sqrt{n}$, provided the sample size n is large (> 30). This result is based on an important theorem in probability called the *central limit theorem*.

Suppose, for example, that the pdf of the random variable X is specified by

$$f(x) = 1/2; \quad 0 < x < 2$$

$$= 0; \quad \text{elsewhere}$$

Application of the equation defining the expected value of X (see Equation 8.2 of Chapter 8) gives

$$\mu = E(X) = \int_0^2 x\left(\frac{1}{2}\right) dx = 1$$

$$E(X^2) = \int_0^2 x^2\left(\frac{1}{2}\right) dx = \frac{4}{3}$$

Therefore, application of the defining equation for the variance (see Equation 8.7 of Chapter 8) yields

$$\sigma^2 = E(X^2) - \mu^2 = \frac{4}{3} - 1 = \frac{1}{3}$$

Thus, if $\overline{X}$ is the mean of a random sample of 48 observations on X, $\overline{X}$ is approximately normally distributed with mean 1 and standard deviation $\sigma/\sqrt{n}$. The latter term is therefore given by

$$\frac{\sigma}{\sqrt{n}} = \frac{\sqrt{1/3}}{\sqrt{48}} = \sqrt{\frac{1}{(3)(48)}} = \sqrt{\frac{1}{144}}$$

$$= \frac{1}{12}$$

The following example is now provided.

$$P(X > 9/8) = P\left[\frac{\overline{X}-1}{1/12} > \frac{9/8-1}{1/12}\right]$$

$$= P(Z > 1.5)$$

$$= 0.067$$

One of the principal applications of the normal distribution in reliability calculations and hazard risk analysis is the distribution of time to failure due to "wear out." Suppose, for example, that a production lot of thermometers, to be employed in an incinerator especially designed to withstand high temperatures and intense vibrations, has just come off the assembly line. A sample of 25 thermometers from the lot is tested under the specified heat and vibration conditions. Time to failure, in hours, is recorded for each of the 25 thermometers. Application of the equations for sample mean X and sample variance s^2 yields

$$\overline{X} = 125; \quad s^2 = 92$$

Past experience indicates that the "wear out" time of this "unit," like that of a large variety of products in many different industries, tends to be normally distributed. Using the previously calculated values of X and s as best estimates of μ and σ, one obtains for the 110-h reliability of the thermometers.

$$P(X > 110) = P\left[\frac{X-125}{\sqrt{92}} > \frac{110-125}{\sqrt{92}}\right]$$

$$= P(Z > -1.56)$$

$$= 0.94$$

As indicated earlier, the normal distribution is symmetric. The data from a normal distribution could be plotted on special graph paper, known as normal probability paper. The resulting plot appears as a straight line. The parameters μ and σ, can be estimated from such a plot. This procedure is demonstrated in Chapter 19. A nonlinear plot on this paper is indicative of nonnormality.

Actual (experimental) data have shown many physical variables to be normally distributed. Examples include physical measurements on living organisms, molecular velocities in an ideal gas, scores on an intelligence test, the average temperatures in a given locality, etc. Other variables, though not normally distributed *per se*, sometimes approximate a normal distribution after an appropriate transformation, such as taking the logarithm (see Chapter 19) or square root of the original variable. The normal distribution also has the advantage that it is tractable mathematically. Consequently, many of the techniques of statistical inference have been derived under the assumption of underlying normal variants.

PROBLEMS AND SOLUTIONS

NOR 1: Normal Distribution Factors

1. Discuss some of the limitations of the normal distribution.
2. Introduce the concept of the level of significance as it applies to normal distribution.

Solution

1. On account of the prominence (and perhaps the name) of the normal distribution, it is sometimes assumed that a random variable is normally distributed, unless proven otherwise. This notion could lead to incorrect results. For example, the normal distribution is generally inappropriate as a model of time to failure. Frequently, a normal distribution provides a reasonable approximation to the main part of the distribution, but is inadequate at one or both tails. Finally, certain phenomena are just not symmetrically distributed, as is required for normality.
   The errors of incorrectly assuming normality depend on the use to which this assumption is applied. Many statistical models and methods derived under this assumption remain valid under moderate deviations from it. On the other hand, if the normality assumption is used to determine the proportion of "items" above or below some extreme limit (e.g., at the tail of the distribution), serious errors might result.

2. If

$$F(z) = \int_0^z \frac{1}{\sqrt{2\pi}} e^{-\frac{z^2}{2}} dz$$

the term

$$1 - 2[1 - F(z)]$$

is referred to as the *confidence level* and provides the probability that the "result" will be in the range $\mu = \pm z$. The term $2[1 - F(z)]$ is denoted by α, e.g.,

$$\alpha = 1 - 2[1 - F(z)]$$

and referred to as the *level of significance*. The term α therefore represents the probability of being in error if one observes a result that is displaced from the mean by $z\sigma$. This "type" of error will be revisited in Section III.

NOR 2: Probability Calculations for Standard Normal Variables

Given a standard normal variable Z, determine $P(0 < Z < 1)$ from a table of areas under a standard normal curve and deduce the following probabilities:

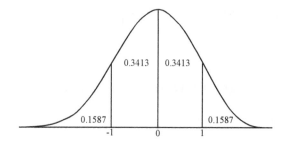

FIGURE 18.5 Normal areas.

1. $P(-1 < Z < 0)$
2. $P(-1 < Z < 1)$
3. $P(Z > -1)$
4. $P(Z < 1)$
5. $P(Z > 1)$

Solution

Obtain $P(0 < Z < 1)$ from Table A.1B (Appendix A). As noted in the Introduction, this table gives the area under a standard normal curve between 0 and various positive values of z. $P(0 < Z < 1)$ is the area between 0 and 1. Therefore,

$$P(0 < Z < 1) = 0.3413$$

Note once again that because the standard normal curve is symmetric about 0, the area to the right of 0 is equal to the area to the left of 0, namely 0.5. Refer to Figure 18.5. Because the area over (0, 1) is 0.3413, the area over (1, ∞) is 0.5 – 0.3413 = 0.1587. By symmetry, the area over (–1, 0) is 0.3413 and the area over (–∞, –1) is 0.1587. These areas are bounded by ordinates erected at –1, 0, and 1, as noted in Figure 18.5.

Additional probabilities are represented by areas under the standard normal curve over certain intervals. From Figure 18.5, one obtains the following additional required probabilities.

1. $P(-1 < Z < 0)$ = Area over (–1, 0) = 0.3413
2. $P(-1 < Z < 1)$ = Area over (–1, 1) = 0.6826
3. $P(Z > -1)$ = Area over (–1, ∞) = 0.3413 + 0.5 = 0.8413
4. $P(Z < 1)$ = Area over (–∞, 1) = 0.5 + 0.3413 = 0.8413
5. $P(Z > 1)$ = Area over (1, ∞) = 0.1587

NOR 3: PROBABILITY CALCULATIONS FOR NONSTANDARD NORMAL VARIABLES

The measurement of the pitch diameter of the thread of a fitting is normally distributed with mean 0.4008 in. and standard deviation 0.0004 in. The specifications are given as 0.4000 ± 0.001 in. The thread diameter meets specifications if the diameter lies between 0.4000 – 0.001 and 0.4000 + 0.001 in. If the thread diameter does not meet specifications, it is considered defective. What is the probability that a defective thread will occur?

Solution

Note that the random variable is not a standard normal variable but is normally distributed. The random variable is the pitch diameter X of the thread of a fitting selected at random from manufactured output. X is normally distributed with mean 0.4008 in. and standard deviation 0.0004 in.

The required probability is the probability that a defect occurs, i.e., that the thread of a fitting produced lies outside the given specification. Therefore, the required probability is the probability that X exceeds $0.4000 + 0.001$ or is less than $0.4000 - 0.001$ in. This required probability is first converted into a probability about a standard normal variable by making use of the fact that if X is normally distributed with mean μ and standard deviation σ and that $Z = (X - \mu)/\sigma$ is a standard normal variable:

$$P(X > 0.4010) + P(X < 0.3990) = 1 - P(0.3990 < X < 0.4010)$$

$$= 1 - P\left[\frac{0.3990 - 0.4008}{0.0004} < \frac{0.4010 - 0.4008}{-0.0004}\right] < \frac{0.4010 - 0.4008}{-0.0004}$$

$$= 1 - P(-4.5 < Z < 0.5)$$

Here, $P(-4.5 < Z < 0.5)$ is the area under a standard normal curve between -4.5 and 0.5.

Proceed to find the area between 0 and 0.5 and then the area between -4.5 and 0. The area between 0 and 0.5 is 0.1915. The area between -4.5 and 0 is the same as the area between 0 and 4.5. The latter area is approximately 0.5. Therefore, the total area between -4.5 and 0.5 is 0.6915, so that

$$P(-4.5 < Z < 0.5) = 0.6915$$

and

$$1 - P(-4.5 < Z < 0.5) = 0.3085$$

Thus, the probability that a defective thread occurs is 0.3085.

NOR 4: TOXIC WASTEWATER

The parts-per-million concentration of a particular toxic substance in a wastewater stream is known to be normally distributed with mean $\mu = 100$ and a standard deviation $\sigma = 2.0$. Calculate the probability that the toxic concentration, C, is between 98 and 104.

Solution

Because C is normally distributed with $\mu = 100$ and a standard deviation $\sigma = 2.0$, then $(C - 100)/2$ is a standard normal variable and

$$P(98 < C < 104) = P\left(-1 < \left[\frac{C - 100}{2}\right] < 2\right)$$

$$= P(-1 < Z < 2)$$

From the values in the standard normal table,

$$P(98 < C < 104) = 0.341 + 0.477$$

$$= 0.818 = 81.8\%$$

NOR 5: ACCEPTANCE LIMITS

Acceptance limits require the plate-to-plate spacing in an electrostatic precipitator to be between 24 and 25 cm. Spacing from other installations suggests that it is normally distributed with an average length of 24.6 cm and a standard deviation of 0.4 cm. What proportion of the plate spacing in a typical unit can be assumed unacceptable?

Solution

The problem is equivalent to determining the probability that an observation or sample from a normal distribution with parameters $\mu = 0$ and $\sigma = 1$ either exceeds

$$Z = \frac{25.0 - 24.6}{0.4} = 1$$

or falls below

$$Z = \frac{24.0 - 24.6}{0.4} = -1.5$$

These probabilities are 0.159 and 0.067, respectively, from a table of the standard normal distribution. Consequently, the proportion of unacceptable spacings is

$$P(\text{unacceptable spacing}) = P(Z > 1) + P(Z < -1.5)$$

$$= 0.159 + 0.067$$

$$= 0.226$$

$$= 22.6\%$$

NOR 6: ESTUARY TEMPERATURE

The temperature of a polluted estuary during the summer months is normally distributed with mean 56°F and standard deviation 3.0°F. Calculate the probability that the temperature is between 55 and 62°F.

Solution

Normalizing the temperature T gives

$$Z_1 = \frac{55 - 56}{3.0} = -0.333$$

$$Z_2 = \frac{62 - 56}{3.0} = 2.0$$

Thus,

$$P(0.333 < Z < 2.0) = P(0.0 < Z < 2.0) - P(0.0 < Z < 0.333)$$

$$= 0.4722 - 0.1293$$

$$= 0.6015$$

$$= 61.15\%$$

NOR 7: TOXIC ASH

The regulatory specification on a toxic substance in a solid waste ash calls for a concentration level of 1.0 ppm or less. Earlier observations of the concentration of the ash, C, indicate a normal distribution with a mean of 0.60 ppm and a standard deviation of 0.20 ppm. Estimate the probability that ash will exceed the regulatory limit.

Solution

This problem requires the calculation of $P(C > 1.0)$. Normalizing the variable C,

$$P\left(\left[\frac{C - 0.6}{0.2}\right] > \left[\frac{1.0 - 0.6}{0.2}\right]\right)$$

$$P(Z > 2.0)$$

From the standard normal table

$$P(Z > 2.0) = 0.0228$$

$$= 2.28\%$$

For this situation, the area to the right of the 2.0 is 2.28% of the total area. This represents the probability that ash will exceed the regulatory limit of 1.0 ppm.

NOR 8: ESP INSTALLED WIRES

The diameter D of wires installed in an electrostatic precipitator (ESP) has a standard deviation of 0.01 in. At what value should the mean be if the probability of its exceeding 0.21 in. is to be 1%?

Solution

For this one-tailed case (with an area of 0.01),

$$P\left(\left[\frac{D - \mu}{\sigma}\right] > \left[\frac{0.21 - \mu}{0.01}\right]\right) = 0.01$$

where

$$Z = \frac{D - \mu}{\sigma}$$

Therefore

$$P\left(Z > \left[\frac{0.21 - \mu}{0.01}\right]\right) = 0.01$$

For a one-tailed test at the 1% (0.01) level,

$$Z = 2.326$$

so that

$$2.326 = \frac{0.21 - \mu}{0.01}$$

$$\mu = 0.187 \text{ in}$$

NOR 9: THERMOMETER LIFETIME VARIANCE

The lifetime, T, of a circuit board employed in an incinerator is normally distributed with a mean of 2500 d. What is the largest lifetime variance the installed circuit boards can have if 95% of them need to last at least 365 d?

Solution

For this application,

$$P(T > 365) = 0.95$$

Normalizing gives

$$P\left(\left[\frac{T - \mu}{\sigma}\right] > \left[\frac{356 - 2500}{\sigma}\right]\right) = 0.95$$

$$P\left(Z > -\frac{2135}{\sigma}\right) = -1.645$$

The following equation must apply for this condition:

$$-\frac{2135}{\sigma} = -1.645$$

Solving,

$$\sigma = 1298 \text{ d}$$

In addition,

$$\sigma^2 = 1.68 \times 10^6 \text{ d}^2$$

NOR 10: BAG FABRIC QUALITY

Let X denote the coded quality of bag fabric used in a particular utility baghouse. Assume that X is normally distributed with mean 10 and standard deviation 2.

1. Find c such that $P(|X - 10| < c) = 0.90$
2. Find k such that $P(X > k) = 0.90$

Solution

1. Because X is normally distributed with mean 10 and standard deviation 2, $(X - 10)/2$ is a standard normal variable. For this two-sided test,

$$P(|X - 10| < c) = P(-c < (X - 10) < c)$$

$$= P\left(\frac{-c}{2} < \frac{X - 10}{2} < \frac{c}{2}\right)$$

$$= P\left(\frac{-c}{2} < Z < \frac{c}{2}\right)$$

$$= 2P\left(0 < Z < \frac{c}{2}\right); \quad \text{due to symmetry}$$

Because

$$P(|X - 10| < c) = 0.90$$

$$2P\left(0 < Z < \frac{c}{2}\right) = 0.90$$

$$P\left(0 < Z < \frac{c}{2}\right) = 0.45$$

For this condition to apply,

$$\frac{c}{2} = 1.645$$

$$c = 3.29$$

2. Based on the problem statement,

$$P(X > k) = 0.90$$

$$P\left(\frac{X - \mu}{\sigma} > \frac{k - 10}{2}\right) = 0.90$$

$$P\left(Z > \frac{k - 10}{2}\right) = 0.90$$

For this equation to be valid

$$\frac{k - 10}{2} = -1.28$$

Solving for k gives

$$k = 7.44$$

NOR 11: PROBABILITY CALCULATIONS CONCERNING THE SAMPLE MEAN

1. For a random sample of 16 observations on a random variable X, normally distributed with mean 101 and standard deviation 4, find $P(\overline{X} > 103)$.
2. For a random sample of 12 observations on a random variable X having a pdf specified by

$$f(x) = 1/2; \quad 0 < x < 2$$

$$= 0; \quad \text{elsewhere}$$

find $P(\overline{X} > 5/4)$.

The reader is referred to the Introduction if problems are encountered in the solution.

Solution

If $\overline{X}$ is the mean of a sample from a normal population with mean μ and standard deviation σ, then $\overline{X}$ is normally distributed with mean μ and standard deviation $\sigma/\sqrt{n}$. If the population sampled is not normal, then $\overline{X}$ is approximately normally distributed with mean μ and standard deviation $\sigma/\sqrt{n}$, provided the sample size n is large, i.e., $n > 30$. This large sample distribution of the sample mean is based on the central limit theorem discussed earlier.

1. Note that $\overline{X}$ is normally distributed with mean 101 and standard deviation $4/\sqrt{16} = 1.0$. Therefore,

$$P(\overline{X} > 101) = P\left(\left[\frac{\overline{X} - 101}{1.0}\right] > \left[\frac{103 - 101}{1.0}\right]\right)$$

$$= P(Z > 2)$$

$$= 0.023$$

2. First compute the mean μ and variance σ^2 of the random variable X:

$$\mu = E(X) = \int_0^2 x(1/2)\,dx = 1$$

and

$$E(X^2) = \int_0^2 x^2(1/2)\,dx = 4/3$$

Therefore,

$$\sigma^2 = E(X^2) - \mu^2$$
$$= 4/3 - 1$$
$$= 1/3$$

Apply the central limit theorem because the sample size in scenario 2 is large. By the central limit theorem, X is approximately normally distributed with mean 1 and standard deviation

$$\sigma = \sqrt{\frac{1}{3(12)}}$$
$$= \sqrt{\frac{1}{36}}$$
$$= 1/6$$

Therefore,

$$P(\bar{X} > 5/4) = P\left(\left[\frac{\bar{X}-1}{1/6}\right] > \left[\frac{5/4-1}{1/6}\right]\right)$$
$$= P(Z > 1.5)$$
$$= 0.067$$

NOR 12: Environmental Engineering Grades

A firm is interested in estimating the national average number grade of environmental students. It desires to estimate the true grade within 1.0 percentage points with 95% confidence. An earlier study of 75 students provided a mean of 88.3 with a standard deviation of 8.6. How many additional student transcripts should be reviewed to obtain the desired accuracy?

Solution

For this case,

$$P\left(\left[\frac{\overline{X}-88.3}{\sigma/\sqrt{n}}\right]<Z<\left[\frac{\overline{X}+88.3}{\sigma/\sqrt{n}}\right]\right)=0.95; \quad Z=\frac{\overline{X}-\mu}{\sigma/\sqrt{n}}$$

For this equality to apply,

$$Z=1.96$$

For a one percentage point difference.

$$\overline{X}-88.3=1.0$$

with

$$Z=1.96=\frac{1.0}{8.6/\sqrt{n}}$$

Solving gives

$$\sqrt{n}=\frac{(1.96)(8.6)}{1.0}$$

$$n=284$$

Therefore, a total of 284 student transcripts need to be examined. An additional 209, i.e., 284 – 75 transcripts are required.

19 LOG. Log-Normal Distribution

INTRODUCTION

A nonnegative random variable X has a log-normal distribution whenever ln X, i.e., the natural logarithm of X, has a normal distribution. The probability distribution function (pdf) of a random variable X having a log-normal distribution is specified by

$$f(x) = \frac{1}{\sqrt{2\pi}\beta} x^{-1} \exp\left[-\frac{(\ln x - \alpha)^2}{2\beta^2}\right]; \ x > 0 \tag{19.1}$$

$$= 0; \quad \text{elsewhere}$$

The mean and variance of a random variable X having a log-normal distribution are given by

$$\mu = e^{\alpha + \beta^2/2} \tag{19.2}$$

$$\sigma^2 = e^{2\alpha + \beta^2}(e^{\beta^2} - 1) \tag{19.3}$$

Figure 19.1 plots the pdf of the log-normal distribution for $\alpha = 0$ and $\beta = 1$. Probabilities concerning random variables having a log-normal distribution can be calculated from the previously employed tables of the normal distribution. If X has a log-normal distribution with parameters α and β, then the ln X has a normal distribution with $\mu = \alpha$ and $\sigma = \beta$. Probabilities concerning X can therefore be converted into equivalent probabilities concerning ln X.

For example, suppose that X has a log-normal distribution with $\alpha = 2$ and $\beta = 0.1$. Then

$$P(6 < X < 8) = P(\ln 6 < \ln X < \ln 8)$$

$$= \left[\frac{\ln 6 - 2}{0.1} < \frac{\ln X - 2}{0.1} < \frac{\ln 8 - 2}{0.1}\right]$$

$$= P(-2.08 < Z < 0.79)$$

$$= (0.5 - 0.019) + (0.5 - 0.215)$$

$$= 0.481 + 0.285$$

$$= 0.78$$

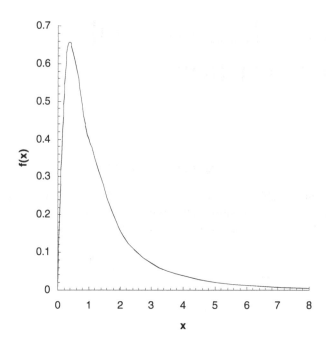

FIGURE 19.1 Log-normal pdf for $\alpha = 0$, $\beta = 1$.

Estimates of the parameters α and β in the pdf of a random variable X having a log-normal distribution can be obtained from a sample of observations on X by making use of the fact that ln X is normally distributed with mean α and standard deviation β. Therefore, the mean and standard deviation of the natural logarithms of the sample observations on X furnish estimates of α and β. To illustrate this procedure, suppose the time to failure T, in thousands of hours, was observed for a sample of five pumps at a water treatment plant. The observed values of T were 8, 11, 16, 22, and 34. The natural logarithms of these observations are 2.08, 2.40, 2.77, 3.09, and 3.53. Assuming that T has a log-normal distribution, the estimates of the parameters α and β in the pdf are obtained from the mean and standard deviation of the natural logs of the observations on T. Applying Equation 9.1 and Equation 9.2 as demonstrated in Problem LOG 1 yields 2.77 as the estimate of α and 0.57 as the estimate of β.

The log-normal distribution has been employed as an appropriate model in a wide variety of situations from environmental management to biology to economics. Additional applications include the distributions of personal incomes, inheritances, bank deposits, and also the distribution of organism growth subject to many small impurities. Perhaps the primary application of the log-normal distribution has been to represent the distribution for particle sizes in gaseous emissions from many industrial processes.

PROBLEMS AND SOLUTIONS

LOG 1: TIME TO FAILURE OF ELECTRIC MOTORS

The time to failure, T, in thousands of hours was observed for a sample of 5 electric motors as follows:

$$8, 11, 16, 22, 34$$

Assuming T has a log-normal distribution estimate the probability that an electric motor lasts less than 5000 h.

Solution

Under the assumption that T has a log-normal distribution, the natural logs of the observation constitute a sample from a normal population. Obtain the natural logs of the given observations on T, time to failure:

$$\ln 8 = 2.08 \quad \ln 16 = 2.77 \quad \ln 34 = 3.53 \quad \ln 11 = 2.40 \quad \ln 22 = 3.09$$

Compute the mean, μ, and the standard deviation, σ, of these results. Employ Equation 9.1 and Equation 9.2,

$$\mu = \sum_{i=1}^{5} \frac{\ln T_i}{5} = 2.77$$

$$\sigma = \sqrt{\sum_{i=1}^{5} \frac{\left[\ln T_i - 2.77\right]^2}{4}} = 0.57$$

Estimate α and β in the pdf of T from μ and σ, respectively.

Estimate of $\alpha = 2.77$

Estimate of $\beta = 0.57$

Convert the required probability concerning T into a probability about $\ln T$:

$$P(T < 5) = P(\ln T < \ln 5) = P(\ln T < 1.61)$$

Treating $\ln T$ as a random variable that is normally distributed with mean α and standard deviation β, obtain the required probability using the standard normal table:

$$P(\ln T < 1.61) = P\left(\left[\frac{\ln T - \alpha}{\beta}\right] < \left[\frac{1.61 - \alpha}{\beta}\right]\right)$$

$$= P\left(Z < \frac{1.61 - 2.77}{0.57}\right)$$

$$= P(Z < -2.04)$$

$$= 0.021$$

$$= 2.1\%$$

LOG 2: MOTOR SURVIVAL TIME

Resolve Problem LOG 1 by calculating the probability that an electric motor lasts more than 5,000 h.

Solution

One can immediately conclude that

$$P(T > 5) = P(\ln T > 1.61) = 1 - P(T < 5)$$

$$= 1.0 - 0.0021$$

$$= 0.979 = 97.9\%$$

LOG 3: EXTENDED MOTOR SURVIVAL TIME

Refer to Problem LOG 1. Calculate the probability that the motor lasts more than 10,000 h.

Solution

The describing equation becomes

$$P(T > 10) = P(\ln T > \ln 10) = P(\ln T > 2.303)$$

Converting to a standard normal variable,

$$P(\ln T > 2.303) = P\left[\left(\frac{\ln T - \alpha}{\beta}\right) > \left(\frac{2.303 - \alpha}{\beta}\right)\right]$$

$$= P\left[Z > \left(\frac{2.303 - 2.77}{0.57}\right)\right]$$

$$= P(Z > -2.434)$$

$$= 0.9926 = 99.26\%$$

LOG 4: COOLANT RECYCLE PUMP

The failure rate per year, Y, of a coolant recycle pump in a wastewater treatment plant has a log-normal distribution. If $\ln Y$ has a mean of 2.0 and variance of 1.5, find $P(0.175 < Y < 1)$.

Solution

If Y has a log-normal distribution, $\ln Y$ has a normal distribution with mean 2 and standard deviation $\sigma = 1.5^{1/2} = 1.22$. Therefore,

$$P(0.175 < Y < 1) = P(\ln 0.175 < \ln Y < \ln 1)$$

$$= P\left(\left[\frac{\ln 0.175 - 2}{1.22}\right] < \frac{\ln Y - 2}{1.22} < \left[\frac{\ln 1 - 2}{1.22}\right]\right)$$

$$= P(-3.07 < Z < -1.64)$$

$$= 0.1587 - 0.0011$$

$$= 0.1576$$

LOG 5: INTERREQUEST TIMES FOR COMPUTER SERVICE

The data in Table 19.1 represent the time in microseconds between requests for a certain process service on a computer network.

TABLE 19.1
Computer Network Data

114462	10280	2654	6761	8111
5437	14691	4605	9405	15184
4866	4789	11944	6919	5547
1439	1333	18270	35632	17783
13017	32145	7310	1812	15078
4138	7361	9405	4277	2592
1594	39577	3820	6925	6974
1422	6063	5432	6003	27778
36938	15615	2904	8840	3711
10829	5575	6634	3674	5825

TABLE 19.2
Computer Network Frequency Distribution

Interrequest Time (microseconds)	Frequency
0–2499	5
2500–4999	11
5000–9999	18
10000–19999	10
20000–39999	5
40000–79999	0
80000–159999	1
Total = 50	Total = 50

Construct a frequency distribution for the data. Assuming that interrequest time has a log-normal distribution, obtain the theoretical frequencies for each class of the frequency distribution.

Solution

A frequency distribution is a table showing the number of observations in each of a succession of intervals called *classes*, as selected in Table 19.2.

Compute the mean, μ, and standard deviation σ, of the natural logs of the observations. Employ the results provided in Table 19.3. The mean, μ, is therefore

$$\mu = \sum_{i=1}^{50} \frac{\ln X_i}{50} = \frac{446.51}{50} = 8.93$$

The standard deviation, is

$$\sigma = \sqrt{\sum_{i=1}^{50} \frac{\left[\ln X_i - 8.93 \right]^2}{49}} = 0.91$$

Estimate α and β by the mean and standard deviation obtained in Table 19.3.

TABLE 19.3
Natural Log Frequency

X	ln X	X	ln X	X	ln X
5437	8.60	6063	8.71	1812	7.5
4866	8.50	15615	9.66	4277	8.36
1439	8.49	5575	8.63	6925	8.84
13017	9.47	2654	7.88	6003	8.70
4138	8.33	4605	8.43	8840	9.09
1594	7.37	11944	9.39	3674	8.21
114462	11.65	18270	9.81	8111	9.00
1422	7.26	7310	8.90	15184	9.63
36938	10.52	9405	9.15	5547	8.62
10829	9.29	3820	8.25	17783	9.79
10280	9.24	5432	8.60	15078	9.62
14691	9.59	2904	7.97	2592	7.86
4789	8.47	6634	8.80	6974	8.85
1333	7.20	6761	8.82	27778	10.23
32145	10.38	9405	9.15	3711	8.22
7361	8.90	6919	8.84	5825	8.67
39577	10.59	35632	10.48		

Estimate of $\alpha = 8.93$

Estimate of $\beta = 0.91$

Convert the probabilities associated with each class of the frequency distribution into corresponding probabilities about the natural logs of the endpoints of each class.

$$P(0 < X < 2499) = P(-\infty < \ln X < \ln 2499) = P(\infty < \ln X < 7.82)$$

$$P(2499 < X < 4999) = P(\ln 2500 < \ln X < \ln 4999) = P(7.82 < \ln X < 8.52)$$

$$P(5000 < X < 9999) = P(\ln 5000 < \ln X < \ln 9999) = P(8.52 < \ln X < 9.21)$$

$$P(10000 < X < 19999) = P(\ln 10000 < \ln X < \ln 19999) = P(9.21 < \ln X < 9.90)$$

$$P(20000 < X < 39999) = P(\ln 20000 < \ln X < \ln 39999) = P(9.90 < \ln X < 10.61)$$

$$P(40000 < X < 79999) = P(\ln 40000 < \ln X < \ln 79999) = P(10.60 < \ln X < 11.29)$$

$$P(80000 < X < 159999) = P(\ln 80000 < \ln X < \ln 159999) = P(11.29 < \ln X < 11.98)$$

Treating ln X as a random variable normally distributed with mean and standard deviation, evaluate the probabilities in the preceding data using the standard normal table.

TABLE 19.4
Interrequest Frequency Results

Interrequest Time (microseconds)	Frequency	Probabilities	Theoretical Frequency
0–2499	5	0.111	5.6
2500–4999	11	0.215	10.8
5000–9999	18	0.296	14.8
10000–19999	10	0.231	11.8
20000–39999	5	0.109	5.5
40000–79999	0	0.028	1.4
80000–159999	1	0.005	0.3

$$P(-\infty < \ln X < 7.82) = P\left(-\infty < \left[\frac{(\ln X - 8.93)}{0.91}\right] < \left[\frac{(7.82 - 8.93)}{0.91}\right]\right)$$

$$= P((-\infty < Z < -1.22)$$

$$= 0.111$$

Similarly,

$$P(7.82 < \ln X < 8.52) = P(-1.22 < Z < 0.45)$$

$$= 0.326 - 0.111$$

$$= 0.215$$

$$P(8.52 < \ln X < 9.21) = P(-0.45 < Z < 0.31)$$

$$= 0.296$$

$$P(9.21 < \ln X < 9.90) = P(0.31 < Z < 1.07)$$

$$= 0.236$$

$$P(9.90 < \ln X < 10.60) = P(1.07 < Z < 1.84)$$

$$= 0.109$$

$$P(10.60 < \ln X < 11.29) = P(1.84 < Z < 2.59)$$

$$= 0.028$$

$$P(11.29 < \ln X < 11.98) = P(2.59 < Z < 3.35)$$

$$= 0.005$$

TABLE 19.5
Particle Size Data

Particle Size Range, d_p, μm	Distribution (μg/m³)
< 0.62	25.5
0.62–1.0	33.15
1.0–1.2	17.85
1.2–3.0	102.0
3.0–8.0	63.75
8.0–10.0	5.1
> 10.0	7.65
Total	255.0

Obtain the theoretical frequency for each class by multiplying each of these probabilities by 50, the total tabulated frequency. Results are presented in Table 19.4.

LOG 6: PARTICLE SIZE DISTRIBUTION DATA

You have been requested to determine if a particle size distribution is log-normal. Data are provided in Table 19.5.

Particulates discharged from an operation, usually to the atmosphere, consist of a size distribution ranging anywhere from extremely small particles (less than 1 μm) to very large particles (greater than 100 μm). Particle size distributions are usually represented by a cumulative weight fraction curve in which the fraction of particles less than or greater than a certain size is plotted against the dimension of the particle.

To facilitate recognition of the size distribution, it is useful to plot a size-frequency curve. The size-frequency curve shows the number (or weight) of particles present for any specified diameter. Because most dusts are comprised of an infinite range of particle sizes, it is first necessary to classify particles according to some consistent pattern. The number or weight of particles may then be defined as that quantity within a specified size range having finite boundaries and typified by some average diameter.

The shape of the curves obtained to describe the particle size distribution generally follows a well-defined form. If the data include a wide range of sizes, it is often better to plot the frequency (i.e., number of particles of a specified size) against the logarithm of the size. In most cases, an asymmetrical or "skewed" distribution exists; normal probability equations do not apply to this distribution. Fortunately, in most instances, the symmetry can be restored if the logarithms of the sizes are substituted for the sizes. The curve is then said to be logarithmic normal (or log-normal) in distribution. Plotting particle diameter vs. cumulative percentage therefore generates cumulative distribution plots described in the preceding text. For log-normal distributions, plots of particle diameter vs. either percent less than stated size (% LTSS) or percent greater than stated size (% GTSS) produce straight lines on log-probability coordinates. The next three problems address this application.

Solution

Cumulative distribution information can be obtained from the calculated results provided in Table 19.6.

The cumulative distribution can be plotted on log-probability graph paper. The cumulative distribution curve is shown in Figure 19.2. Because a straight line is obtained on log-normal coordinates, the particle size distribution is log-normal.

TABLE 19.6
Cumulative Distribution Data

Particle Size Range, d_p (μm)	Total (%)	Cumulative % GTSS
< 0.62	10	90
0.62–1.0	13	77
1.0–1.2	7	70
1.2–3.0	40	30
3.0–8.0	25	5
8.0–10.0	2	3
> 10.0	3	0

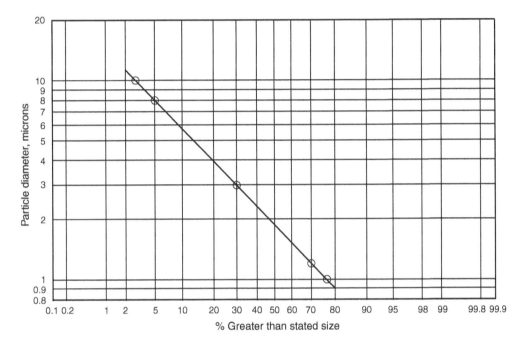

FIGURE 19.2 Cumulative size distribution for Problem LOG 6.

LOG 7: MEAN AND STANDARD DEVIATION OF A PARTICLE SIZE DISTRIBUTION

Refer to Problem LOG 6. Estimate the mean and standard deviation from the size distribution information available.

Solution

The use of probability plots is of value when the arithmetic or geometric mean is required because these values may be read directly from the 50% point on a logarithmic probability plot. By definition, the size corresponding to the 50% point on the probability scale is the *geometric mean diameter*. The geometric standard deviation is given (for % LTSS) by

$$\sigma = 84.14\% \text{ size}/50\% \text{ size}$$

or

$$\sigma = 50\% \text{ size}/15.87\% \text{ size}$$

For % GTSS,

$$\sigma = 50\% \text{ size}/84.14\% \text{ size}$$

or

$$\sigma = 15.87\% \text{ size}/50\% \text{ size}$$

The mean, as read from the 50% GTSS point on the graph in Figure 19.3 is approximately 1.9 μm. A value of 1.91 μm is obtained from an expanded plot.

From the diagram, the particle size corresponding to the 15.87% point is

$$d_p (15.87\%) = 4.66 \ \mu m$$

The standard deviation may now be calculated. By definition,

$$\sigma = d_p (15.87\%)/d_p (50\%)$$

$$= 4.66/1.91$$

$$= 2.44$$

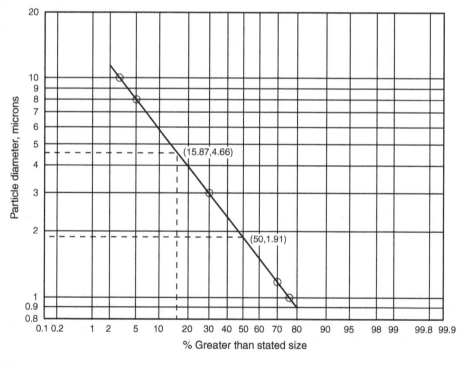

FIGURE 19.3 Cumulative distribution curve for Problem LOG 7.

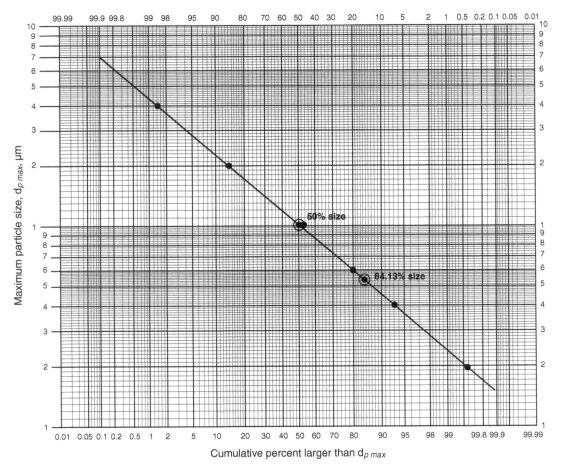

FIGURE 19.4 Log-probability distribution for data from Problem LOG 8.

LOG 8: CUMULATIVE PARTICLE SIZE DATA

The following cumulative particle size data is provided in Table 19.7. Estimate the mean and standard deviation.

Solution

A plot (see Figure 19.4) on log-probability paper yields a straight line. Therefore, the distribution is again log-normal. To determine the geometric mean diameter one can read the 50% size:

$$d_p (50\%) = 10.5 \ \mu m$$

The standard deviation may now be calculated. By definition,

$$\sigma = d_p (50\%)/d_p (84.13\%)$$

$$= 10.5/5.5$$

$$= 1.9$$

TABLE 19.7
Particle Size Concentration Data

Particle Size Range (μm)	Concentration (μg/m³)	Weight in Size Range (%)	Cumulative % GTSS[a]
0–2	0.8	0.4	99.6
2–4	12.2	6.1	93.5
4–6	25.0	12.5	81.0
6–10	56	28	53.0
10–20	76	38	15.0
20–40	27	13.5	1.5
> 40	3	1.5	—

[a] % GTSS represents the percent greater than stated size, where the stated size is the upper limit of the corresponding particle size range. Thus, 99.6% of the particles have a size equal to or greater than 2 μm.

TABLE 19.8
Anderson 2000 Sampler Data

Plate Number	Tare Weight (g)	Final Weight (g)
0	20.48484	20.48628
1	21.38338	21.38394
2	21.92025	21.92066
3	21.55775	21.55817
4	11.40815	11.40854
5	11.61862	11.61961
6	11.76540	11.76664
7	20.99617	20.99737
Backup filter	0.20810	0.21156

LOG 9: ANDERSON 2000 SAMPLER

Given Anderson 2000 sampler data from an oil-fired boiler, an environmental student has been requested to plot a cumulative distribution curve on log-probability paper and determine mean particle diameter and geometric standard deviation of the fly ash. Pertinent data are provided in Table 19.8.
Note that

$$\text{Sampler volumetric flow rate, } q = 0.5 \text{ cfm}$$

See also Figure 19.5 for aerodynamic diameter vs. flow rate data for an Anderson sampler.

Solution

Table 19.9 provides the net weight and percentage of total weight. A sample calculation (for plate 0) follows:

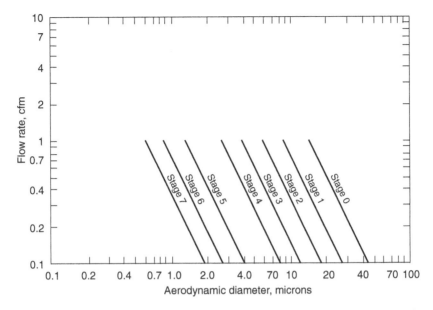

FIGURE 19.5 Aerodynamic diameter vs. flow rate through Anderson sampler for an impaction efficiency of 95%.

TABLE 19.9
Anderson 2000 Sampler Calculations

Plate Number	Net Weight (mg)	Percentage of Total Weight
0	1.44	14.2
1	0.56	5.5
2	0.41	4.1
3	0.42	4.2
4	0.39	3.9
5	0.99	9.8
6	1.24	12.3
7	1.20	11.9
Backup filter	3.46	34.2
Total	10.11	100.0

Net weight = Final weight − Tare weight

$$= 20.48628 - 20.48484$$

$$= 1.44 \times 10^{-3} \text{ g}$$

$$= 1.44 \text{ mg}$$

Percentage of total weight = (net weight/total net weight)(100)

$$= (1.44/10.11)(100\%)$$

$$= 14.2\%$$

TABLE 19.10
Anderson Sampler Cumulative
Percentage Data

Plate No.	Cumulative Percentage
0	85.8
1	80.3
2	76.2
3	72.0
4	68.1
5	58.3
6	46.0
7	34.1
Backup filter	—

Calculate the cumulative percentage for each plate. Again for plate 0,

$$\text{Cumulative \%} = 100 - 14.2$$

$$= 85.8\%$$

For plate 1,

$$\text{Cumulative \%} = 100 - (14.2 + 5.5)$$

$$= 80.3\%$$

Table 19.10 shows the cumulative percentage for each plate.

Using the Anderson graph shown in Figure 19.5, determine the 95% aerodynamic diameter at $q = 0.5$ cfm for each plate (stage). The cumulative distribution curve is provided on log-probability coordinates in Figure 19.6.

Table 19.11 shows the 95% aerodynamic diameter for each plate.

The mean particle diameter is the particle diameter, Y, corresponding to a cumulative percentage of 50.

$$\text{Mean particle diameter} = Y_{50} = 1.6 \ \mu m$$

The distribution appears to approach log-normal behavior. The particle diameter at a cumulative percentage of 84.13 is

$$Y_{84.13} = 15^+ \ \mu m \approx 15 \ \mu m$$

Therefore, the geometric standard deviation is approximately

$$\sigma_G = Y_{84.13} / Y_{50}$$

$$= 15 / 1.6$$

$$= 9.4$$

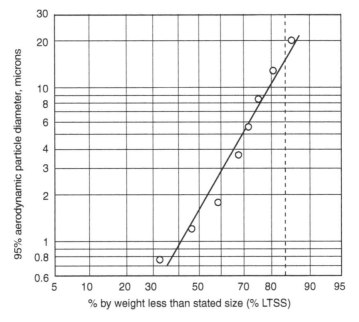

FIGURE 19.6 % LTSS for Problem LOG 9.

TABLE 19.11
Anderson Sampler Aerodynamic Data

Plate No.	95% Aerodynamic Diameter (μm)
0	20.0
1	13.0
2	8.5
3	5.7
4	3.7
5	1.8
6	1.2
7	0.78

LOG 10: BOD Levels

Normalized biological oxygen demand (BOD) levels in an estuary during the past 10 years are summarized in Table 19.12. If the BOD levels are assumed to follow a log-normal distribution, predict the level that would be exceeded only once in 100 years.

TABLE 19.12
Estuary BOD Data

Year	1	2	3	4	5	6	7	8	9	10
BOD Level	23	38	17	210	62	142	43	29	71	31

Note: BOD = biological oxygen demand.

TABLE 19.13
BOD Calculations

Year (Y)	BOD Level	$X = \ln$ BOD	X^2
1	23	3.13	9.83
2	38	3.64	13.25
			8.01
3	17	2.83	8.01
4	210	5.35	28.62
5	62	4.13	17.06
6	142	4.96	24.60
7	43	3.76	14.14
8	29	3.37	11.36
9	71	4.26	18.15
10	31	3.43	11.76
Total	—	38.86	156.78

Solution

For this case, refer to Table 19.13.

Based on the data presented in Table 19.13

$$\bar{X} = \frac{\sum X}{n} = \frac{38.86}{10} = 3.886$$

$$s^2 = \frac{\sum X^2 - \left(\sum X\right)^{2/n}}{n-1}$$

$$= \frac{156.78 - (38.86)^2 / 10}{10-1}$$

$$= 0.64$$

and

$$s = 0.80$$

For this test, with $Z = 2.327$ for the 99% value,

$$Z = \frac{X - \bar{X}}{s}$$

$$2.327 = \frac{X - 3.886}{0.80}$$

Solving for X yields

$$X = 5.748$$

For this log-normal distribution,

$$X = \ln(\text{BOD})$$

$$X = 5.748 = \ln(\text{BOD})$$

$$\text{BOD} = 313$$

20 MON. Monte Carlo Simulation

INTRODUCTION

Monte Carlo simulation is a procedure for mimicking observations on a random variable that permits verification of results that would ordinarily require difficult mathematical calculations or extensive experimentation. The method normally uses computer programs called *random number generators*. A random number is a number selected from the interval (0,1) in such a way that the probabilities that the number comes from any two subintervals of equal "length" are equal. For example, the probability that the number is in the subinterval (0.1,0.3) is the same as the probability that the number is in the subinterval (0.5,0.7). Thus, random numbers are observations on a random variable X having a uniform distribution on the interval (0,1). This means that the probability distribution function (pdf) of X is specified by

$$f(x) = 1; \quad 0 < x < 1$$
$$= 0; \quad \text{elsewhere}$$

(20.1)

The above pdf assigns equal probability to subintervals of equal length in the interval (0,1). Using random number generators, Monte Carlo simulation can generate observed values of a random variable having any specified pdf. For example, to generate observed values of T, the time to failure, when T is assumed to have a pdf specified by $f(t)$, one first uses the random number generator to generate a value of X between 0 and 1. The solution is an observed value of the random variable T having a pdf specified by $f(t)$.

PROBLEMS AND SOLUTIONS

MON 1: MONTE CARLO SIMULATION OF TIME TO FAILURE

A pump has time to failure, T, measured in years, with an exponential pdf specified by

$$f(t) = e^{-t}; \quad t > 0$$
$$= 0; \quad \text{elsewhere}$$

Generate 15 simulated values of T using a Monte Carlo procedure, i.e., the generation of random numbers.

Solution

Calculate the cumulative distribution function (cdf), $F(t)$, from the given pdf specified by $f(t)$.

$$f(t) = e^{-t}; \quad t > 0$$
$$= 0; \quad t \leq 0$$

Applying Equation 7.4, one may write

$$F(t) = P(T \leq t) = \int_0^t f(t)\,dt$$

$$= \int_0^t e^{-t}\,dt$$

$$= 1 - e^{-t}; \quad t > 0$$

$$= 0; \quad t \leq 0$$

Fifteen random numbers are now generated in the interval or range of 0 to 1:

0.93, 0.06, 0.53, 0.56, 0.41, 0.38, 0.78, 0.54,

0.49, 0.89, 0.77, 0.85, 0.17, 0.34, 0.56

For each random number generated, solve the equation obtained by setting the random number equal to $F(t)$. The first random number generated is 0.93. Setting this equal to the cdf, $F(t)$, produces the equation

$$0.93 = 1 - e^{-t}$$

Solving yields

$$e^{-t} = 0.07$$

$$t = -\ln 0.07 = 2.66$$

Therefore, the first simulated value of T, time to failure, is 2.66 years. The other simulated values of T obtained in the same manner are shown in Table 20.1.

MON 2: AVERAGE TIME TO FAILURE

Use the results of Problem MON 1 to estimate the average life of the pump.

Solution

The average value of the 15 simulated values of T obtained by dividing the sum by 15 is 1.02 years; this represents the Monte Carlo estimate of the average life of the pump.

MON 3: OTHER PREDICTIVE APPROACHES

Outline at least two other methods that could be employed to estimate the average life of the pump in Problem MON 1.

**TABLE 20.1
Simulated Time to Failure**

Random Number	Simulated Time to Failure (in years)
0.93	2.66
0.06	0.06
0.53	0.76
0.41	0.82
0.38	0.48
0.78	1.52
0.54	0.78
0.49	0.67
0.89	2.21
0.77	1.47
0.85	1.90
0.17	0.19
0.34	0.42
0.56	0.82

Solution

The exact value of the average life of the pump can be calculated by finding the expected value of T from its pdf. Applying Equation 8.2 gives

$$E(T) = \int_{-\infty}^{\infty} t f(t)\, dt$$

$$= \int_{0}^{\infty} t e^{-t}\, dt$$

$$= 1.0$$

A more accurate estimate of the true value of the average life of the pump can be obtained by increasing the number of simulated values on which the estimate is based. The expected value, or mean, can be shown to equal the coefficient associated with the exponential term, i.e., $(1)t$, for an exponential distribution. Therefore, the mean for this distribution is (1).

Generally, as indicated earlier, Monte Carlo simulation can also be used to estimate some function of one or more random variables when direct calculation of the function would be difficult, if not impossible.

MON 4: Monte Carlo Simulation of Normally Distributed Times to Failure

A series system consists of two electrical components, A and B. Component A has a time to failure, T_A, assumed to be normally distributed with mean (μ) 100 h and standard deviation (σ) 20 h. Component B has a time to failure, T_B, assumed to be normally distributed with mean 90 h and standard deviation 10 h. The system fails whenever either component A or component B fails. Therefore, T_S, the time to failure of the system is the minimum of the times to failure of components A and B.

TABLE 20.2
Simulated Values of Z

Random Number	Simulated Values of Z
0.10	−1.28
0.54	0.10
0.42	−0.20
0.02	−2.05
0.81	0.88
0.07	−1.48
0.06	−1.56
0.27	−0.61
0.57	0.18
0.80	0.84
0.92	1.41
0.86	1.08
0.45	−0.13
0.38	−0.31
0.88	1.17
0.21	−0.81
0.26	−0.64
0.51	0.03
0.73	0.61
0.71	0.56

Estimate the average value of T_S on the basis of the simulated values of 10 simulated values of T_A and 10 simulated values of T_B.

Solution

First, generate 20 random numbers in the range of 0 to 1:

$$0.10, 0.54, 0.42, 0.02, 0.81, 0.07, 0.06, 0.27, 0.57, 0.80,$$

$$0.92, 0.86, 0.45, 0.38, 0.88, 0.21, 0.26, 0.51, 0.73, 0.71$$

Use the table of the standard normal distribution (see Table A.1A of Appendix A) and obtain the simulated value of Z corresponding to each of the random numbers. The first random number is 0.10. The corresponding simulated value of Z is −1.28 because the area under a standard normal curve to the left of −1.28 is 0.10. The remaining simulated values of Z are obtained in similar fashion. The 20 simulated values of Z are provided in Table 20.2.

Using the first 10 simulated values of Z, obtain 10 simulated values of T_A by multiplying each simulated value of Z by 20 (the standard deviation σ) and adding 100 (the mean, μ), i.e.,

$$T_A = \sigma Z + 100$$

Since

$$Z = \frac{T_A - 100}{\sigma}$$

TABLE 20.3
Minimum Simulated Values

Simulated Time to Failure		
Component A (T_A)	Component B (T_B)	System (T_S)
74	104	74
102	101	101
96	89	89
59	87	59
118	102	102
70	82	70
69	84	69
88	90	88
104	96	96
117	96	96

Note that the lifetime or time to failure of each component, T, is calculated using this equation. Thus, multiplying each of the first 10 simulated values of Z by 20 and adding 100 yields the following simulated values of T_A:

$$74, 102, 96, 59, 118, 70, 69, 88, 104, 117$$

Multiplying each of the second 10 simulated values of Z by 10 and adding 90 yields the following simulated values of T_B:

$$104, 101, 89, 87, 102, 82, 84, 90, 96, 96$$

Obtain simulated values of T_S corresponding to each pair of simulated values of T_A and T_B are obtained by recording the minimum of each pair. The values are shown in Table 20.3.

The average of the 10 simulated values of T_S is 84, the estimated time to failure of the system.

MON 5: AFTERBURNER THERMOMETER LIFETIME

According to state regulations, three thermometers (A, B, C) are positioned near the outlet of an afterburner. Assume that the individual thermometer component lifetimes are normally distributed with means and standard deviations given in Table 20.4.

Using the 10 random numbers from 0 to 1 provided in Table 20.5 for each thermometer, simulate the lifetime (time to thermometer failure) of the temperature recording system, and estimate its mean and standard deviation, and the estimated time to failure for this system. The lifetime is defined as the time (in weeks) for one of the thermometers to "fail."

TABLE 20.4
Thermometer Data

Thermometer	A	B	C
Mean (weeks)	100	90	80
Standard deviation (weeks)	30	20	10

TABLE 20.5
Thermometer Random Numbers

For A		For B		For C	
0.52	0.01	0.77	0.67	0.14	0.90
0.80	0.50	0.54	0.31	0.39	0.28
0.45	0.29	0.96	0.34	0.06	0.51
0.68	0.34	0.02	0.00	0.86	0.56
0.59	0.46	0.73	0.48	0.87	0.82

Solution

Let T_A, T_B, and T_C denote the lifetimes of thermometer components A, B, and C, respectively. Let T_S denote the lifetime of the system. The random number generated in Table 20.5 may be viewed as the cumulative probability, and the cumulative probability is the area under the standard normal distribution curve. Because the standard normal distribution curve is symmetrical, the negative values of Z and the corresponding area are once again found by symmetry. For example, as described earlier,

$$P(Z < -1.54) = 0.062$$

$$P(Z > 1.54) = 0.062$$

$$P(0 < Z < 1.54) = 0.5 - P(Z > 1.54)$$

$$= 0.5 - 0.062$$

$$= 0.438$$

Recall that the lifetime or time to failure of each component, T, is calculated using the equation

$$T = \mu + \sigma Z$$

where μ is the mean, σ, the standard deviation, and Z, the standard normal variable.

First, determine the values of the standard normal variable, Z, for component A using the 10 random numbers given in the problem statement and the standard normal table. Then calculate the lifetime of thermometer component A, T_A, using the equation for T, given in the preceding text (see Table 20.6A).

Next, determine the values of the standard normal variable and the lifetime of the thermometer component for component B (see Table 20.6B).

Also, determine the values of the standard normal variable and the lifetime of the thermometer component for component C (see Table 20.6C).

For each random value of each component, determine the system lifetime, T_S. Because this is a series system, the system lifetime is limited by the component with the minimum lifetime (see Table 20.7).

The mean value, μ, of T_S is

$$\mu = \frac{635}{10} = 63.5 \text{ years}$$

TABLE 20.6A
Lifetime of Thermometer A (T_A)

Random Number	Z (from standard normal table)	$T_A = 100 + 30 Z$
0.52	0.05	102
0.80	0.84	125
0.45	−0.13	96
0.68	0.47	114
0.59	0.23	107
0.01	−2.33	30
0.50	0.00	100
0.29	−0.55	84
0.34	−0.41	88
0.46	0.10	97

TABLE 20.6B
Lifetime of Thermometer B (T_B)

Random Number	Z (from standard normal table)	$T_B = 90 + 20 Z$
0.77	0.74	105
0.54	0.10	92
0.96	1.75	125
0.02	−2.05	49
0.73	0.61	102
0.67	0.44	99
0.31	−0.50	80
0.34	−0.41	82
0.00	−3.90	12
0.48	−0.05	89

TABLE 20.6C
Lifetime of Thermometer C (T_C)

Random Number	Z (from standard normal table)	$T_C = 80 + 10 Z$
0.14	−1.08	69
0.39	−0.28	77
0.06	−1.56	64
0.86	1.08	91
0.87	1.13	91
0.90	1.28	93
0.28	−0.58	74
0.51	0.03	80
0.56	0.15	81
0.82	0.92	89

TABLE 20.7
Thermometer System Lifetime

T_A	T_B	T_C	T_S
102	105	69	69
125	92	77	77
96	125	64	64
114	49	49	49
107	102	91	91
30	99	30	30
100	80	74	74
84	82	80	80
88	12	12	12
97	89	89	89
Total			635

Calculate the standard deviation, σ, of T_S using the equation

$$\sigma^2 = \frac{1}{n}\sum (T_S - \mu)^2$$

where n is 10, the number in the population. Note that this is not a sample, so that a modified form of Equation 9.2 applies for σ^2 (see Table 20.8).

Therefore,

$$\sigma = \left(\frac{5987}{10}\right)^{0.5} = 24.5 \text{ years}$$

Monte Carlo simulation is an extremely powerful tool available to the scientist/engineer that can be used to solve multivariable systems, ordinary and partial differential equations, numerical integrations, etc. The application to solving differential equations is provided in Problem MON 6.

TABLE 20.8
Thermometer Standard
Deviation Calculations

System Lifetime (T_S)	$(T_S \mu)^2$
69	30.25
77	182.25
64	0.25
49	210.25
91	756.25
30	1122.25
74	110.25
80	272.25
12	2652.25
89	650.25
Total	5987

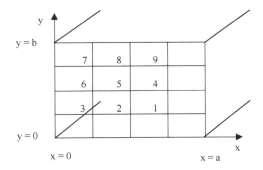

FIGURE 20.1 Monte Carlo grid (square surface).

MON 6: HEAT CONDUCTION: RECTANGULAR PARALLELEPIPED (SQUARE SURFACE)

A new radiant heat transfer rod has been proposed for use in incineration. One of the first steps in determining the usefulness of this form of "hot" rod, is to analytically estimate the temperature profile in the rod.

A recent environmental engineering graduate has proposed to estimate the temperature profile of the square pictured in Figure 20.1. She sets out to outline a calculational procedure to determine the temperature profile. One of the options available is to employ the Monte Carlo method. Provide an outline of that procedure.

Solution

One method of solution involves the use of the Monte Carlo approach, involving the use of random numbers. Consider the square (it could also be a rectangle) pictured in Figure 20.1. If the describing equation for the variation of T within the grid structure is

$$\frac{\partial^2 T}{\partial x^2} + \frac{\partial^2 T}{\partial y^2} = 0$$

with specified boundary conditions (BC) for $T(x, y)$ of T(0, y), T(a, y), T(x, 0), and T(x, b), one may employ the following approach.

1. Proceed to calculate T at point 1, i.e., T_1.
2. Generate a random number between 00 and 99.
3. If the number is between 00 and 24, move to the left. For 25 to 49, 50 to 74, and 75 to 99, move upward, to the right, and downward, respectively.
4. If the move in step 3 results in a new position that is at an outer surface (boundary), terminate the first calculation for point 1 and record the T value of the boundary at the new position. However, if the move results in a new position that is not at a boundary and is still at one of the nine internal grid points, repeat step 2 and step 3. This process is continued until an outer surface or boundary is reached.
5. Repeat step 2 to step 4 numerous times; e.g., 1000 times.
6. After completing step 5, sum all the T values obtained and divide this value by the number of times step 2 to step 4 have been repeated. The resulting value provides a reasonable estimate of T_1.
7. Return to step 1 and repeat the calculations for the remaining 8 grid points.

As mentioned earlier, this method of solution is not limited to square systems.

20	19	18	17	16	
11	12	13	14	15	
10	9	8	7	6	
1	2	3	4	5	

FIGURE 20.2 Rectangular grid.

MON 7: HEAT CONDUCTION: RECTANGULAR PARALLELEPIPED (RECTANGULAR SURFACE)

Outline how to solve Problem MON 6 if the face surface is a rectangle.

Solution

Consider the grid for the face surface in Figure 20.2.
    The outline of the calculation presented in Problem MON 6 remains the same.

MON 8: HEAT CONDUCTION: HOLLOW PIPE

Consider the system provided in Figure 20.3.

Solution

A "grid" of several annular circles must be set at points that produce equal annular areas. Unlike the previous two problems, this involves a one-dimensional calculation. The procedure essentially remains the same for determining the temperature at each grid. The analytical solution is simply given by (personal notes: L. Theodore)

$$T = T_a + (T_b - T_a)\left(\frac{b}{b-a}\right)\left(1 - \frac{a}{r}\right); \quad a = \text{inner radius}$$

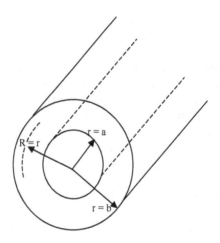

FIGURE 20.3 Hollow pipe grid.

21 FET. Fault Tree and Event Tree Analysis

INTRODUCTION

A *fault tree* is a graphical technique used to analyze complex systems. The objective is to spotlight faulty conditions that cause a system to fail. Fault tree analysis attempts to describe how and why an accident or other undesirable events has occurred. It may also be used to describe how and why an accident or other undesirable event could take place. A fault tree analysis also finds wide application in environmental management as it applies to hazard analysis and risk assessment of process and plant systems.

Fault tree analysis seeks to relate the occurrence of an undesired event, the "top event," to one or more antecedent events, called *basic events*. The top event may be, and usually is, related to the basic events via certain intermediated events. A fault tree diagram exhibits the causal chain linking the basic events to the intermediate events and the latter to the top event. In this chain, the logical connection between events is indicated by so-called *logic gates*. The principal logic gates are the "AND" gate, symbolized on the fault tree by ⌂, and the "OR" gate symbolized by ⌂.

As a simple example of a fault tree, consider a water pumping system consisting of two pumps A and B, where A is the pump ordinarily operating and B is a standby unit that automatically takes over if A fails. A control valve in both cases regulates flow of water through the pump. Suppose that the top event is no water flow, resulting from the following basic events: failure of pump A and failure of pump B or failure of the control valve. The fault tree diagram for this system is shown in Figure 21.1.

Because one of the purposes of a fault tree analysis is the calculation of the probability of the top event, let A, B, and C represent the failure of pump A, the failure of pump B, and the failure of the control valve, respectively. Then, if T represents the top event, no water flow, one can write

$$T = AB + C$$

This equation indicates that T occurs if both A and B occur or if C occurs.

Assume that A, B, and C are independent and $P(A) = 0.01$, $P(B) = 0.005$, and $P(C) = 0.02$. Then, application of the addition theorem in Equation 4.6 in Chapter 4 yields

$$P(T) = P(AB) + P(C) - P(ABC)$$

The independence of A, B, and C implies

$$P(AB) = P(A)P(B)$$

and

$$P(ABC) = P(A)\ P(B)\ P(C)$$

211

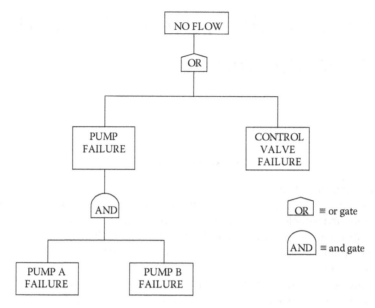

FIGURE 21.1 Fault tree for a water pumping system.

Therefore, the preceding equation can be written as

$$P(T) = P(A)\,P(B) + P(C) - P(A)\,P(B)\,P(C) \qquad (21.1)$$

Substituting the preceding data gives

$$P(T) = (0.01)(0.005) + 0.02 - (0.01)(0.005)(0.02)$$

$$= 0.020049$$

In connection with fault trees, cut sets and minimal cut sets are defined as follows. A *cut set* is a basic event or intersection of basic events that will cause the top event to occur. A *minimal cut set* is a cut set that is not a subset of any other cut set. In the example represented in the preceding text, AB and C are cut sets because if AB occurs then T occurs, and if C occurs, then T occurs. Also, AB and C are minimal cut sets, because neither is a subset of the other. The event T described can be regarded as a top event represented as a union of cut sets.

Consider now Equation 21.2

$$T = AE + AEF + BE + BEF \qquad (21.2)$$

Here, AE, AEF, BE, and BEF are cut sets, because the occurrence of any one of the corresponding events results in the occurrence of T. AEF is not a minimal cut set because it is a subset of AE. Also, BEF is not a minimal cut set because it is a subset of BE. The minimal cut sets in this case are, therefore, AE and BE.

Consider the following hypothetical example of a more complicated fault tree involving eight basic events leading to the rupture of a pressure release disk, the top event T. The fault tree is shown in Figure 21.2. Let basic events be defined as follows:

      A   Premature disk failure
      B   Operator error

C   Pump motor failure
D   Reaction inhibitor system failure
E   Coolant system failure
F   Outlet piping obstruction
G   Motor alarm failure
H   Pressure sensor failure

The top event T can be represented in terms of the basic events as follows:

$$T = A + C\ (G + H) + B + DE + (F + G) \qquad (21.3)$$

Applying the associative law and distributive law of Boolean algebra, given in Equation 3.3 and Equation 3.5 (Chapter 3), to Equation 21.3 yields

$$T = A + CG + CH + B + DE + F + G \qquad (21.4)$$

The cut sets for this system are then

$$A$$
$$CG$$
$$CH$$
$$B$$
$$DE$$
$$F$$
$$G$$

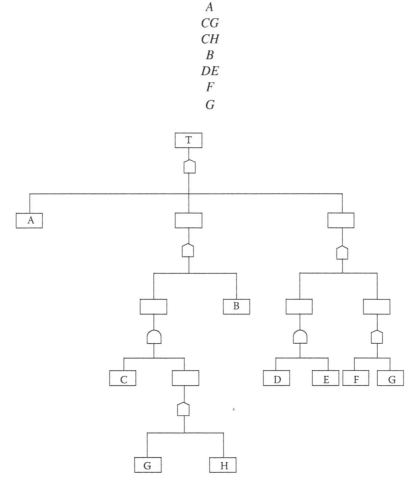

FIGURE 21.2 Fault tree with nine basic events.

The only cut set that is a subset of another cut set is *CG*, which is a subset of *G*. Eliminating *CG* yields the following minimal cut sets:

$$A$$
$$CH$$
$$B$$
$$DE$$
$$F$$
$$G$$

The probability of the top set *T* is the probability of the union of the events represented by the minimal cut sets. This probability usually can be approximated by the sum of the probabilities of these events. Suppose that all the basic events are independent and that their probabilities are

$$P(A) = 0.030$$
$$P(B) = 0.010$$
$$P(C) = 0.005$$
$$P(D) = 0.040$$
$$P(E) = 0.009$$
$$P(F) = 0.070$$
$$P(G) = 0.020$$
$$P(H) = 0.050$$

Then, the probability of the disk rupturing can be approximated by

$$P(T) = P(A) + P(CH) + P(B) + P(DE) + P(F) + P(G)$$

$$= 0.03 + (0.005)(0.05) + 0.01 + (0.04)(0.009) + 0.07 + 0.02$$

$$= 0.1306$$

An *event tree* provides a diagrammatic representation of event sequences that begin with a so-called initiating event and terminate in one or more undesirable consequences. In contrast to a fault tree, which works backward from an undesirable consequence to possible causes, an event tree works forward from the initiating event to possible undesirable consequences. The initiating event may be equipment failure, human error, power failure, or some other event that has the potential for adversely affecting an ongoing process or environment.

The following illustration of event tree analysis is based on one reported by Lees, in his classic work [1]. Consider a situation in which the probability of an external power outage in any given year is 0.1. A backup generator is available, and the probability that it will fail to start on any given occasion is 0.02. If the generator starts, the probability that it will not supply sufficient power for the duration of the external power outage is 0.001. An emergency battery power supply is available; the probability that it will be inadequate is 0.01.

Figure 21.3 shows the event tree with the initiating event, external power outage, denoted by *I*. Labels for the other events on the event tree are also indicated. The event sequences *I S G B* and *I S B* terminate in the failure of emergency power supply. Applying the multiplication theorem in the form given in Equation 5.8 (see Chapter 5), one obtains

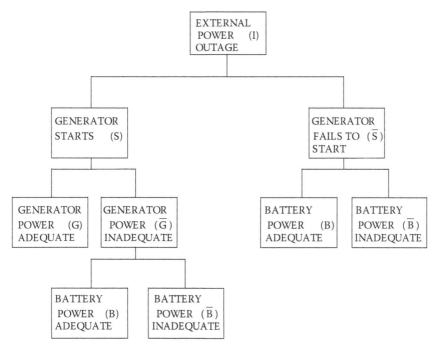

FIGURE 21.3 Event tree for a power outage.

$$P(IS\overline{G}\overline{B}) = (0.1)(0.98)(0.001)(0.01)$$

$$= 0.098 \times 10^{-5}$$

$$P(I\overline{S}\overline{B}) = (0.1)(0.02)(0.01)$$

$$= 2 \times 10^{-5}$$

Therefore, the probability of emergency power supply failure in any given year is 2.098×10^{-5}, the sum of these two probabilities.

PROBLEMS AND SOLUTIONS

FET 1: DECISION TREES

Construct a decision tree given the following information:

Date: a couple's anniversary
Decision: buy flowers or do not buy flowers
Consequences (buy flowers): domestic bliss or suspicious wife
Consequences (do not buy flowers): status quo, or wife in tears, or husband in doghouse

Solution

Begin by setting the initial event in "tree" format which is the top event (Figure 21.4). Also note that this is an event tree. Set up the first branch from the initial event. This is the first decision

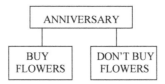

FIGURE 21.4 Top event.

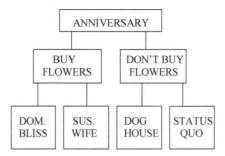

FIGURE 21.5 First decision point.

FIGURE 21.6 Marital bliss event tree application.

point (Figure 21.5). Complete the tree. Note that the bottom four events evolve from what may be defined as resolution of uncertainty points (Figure 21.6).

This is an example of an event tree. As indicated above, in contrast to a fault tree, which works backward from a consequence to possible causes, an event tree works forward from the initiating (or top) event to all possible consequences. Thus, this type of tree provides a diagrammatic representation of sequences that begin with a so-called initiating event and terminate in one or more consequences. It primarily finds application in hazard analysis.

FET 2: PLANT FIRE

If a plant fire occurs, a smoke alarm sounds with probability 0.9. The sprinkler system functions with probability 0.7 whether or not the smoke alarm sounds. The consequences are minor fire damage (alarm sounds, sprinkler works), moderate fire damage with few injuries (alarm fails, sprinkler works), moderate fire damage with many injuries, and major fire damage with many injuries (alarm fails, sprinkler fails). Construct an event tree, and indicate the probabilities for each of the four consequences.

Solution

The first set of consequences of the plant fire with their probabilities are shown in Figure 21.7.

The second set of consequences of the plant fire and their probabilities are shown in Figure 21.8.

The final set of consequences and the probabilities of minor fire damage, moderate fire damage with few injuries, moderate fire damage with many injuries, and major fire damage with many injuries are shown in Figure 21.9.

Note that for each branch in an event tree, the sum of probabilities must equal 1.0. Note again the behavior of an event tree, including the following:

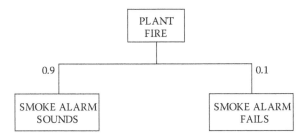

FIGURE 21.7 Event tree with first set of consequences.

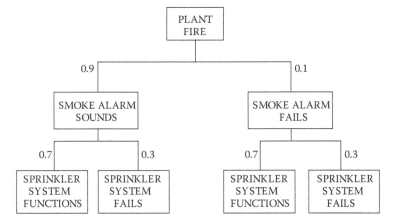

FIGURE 21.8 Event tree with second set of consequences.

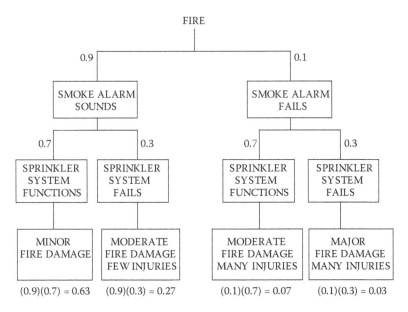

FIGURE 21.9 Event tree with final set of consequences.

1. An event tree works forward from an initial event, or an event that has the potential for adversely affecting an ongoing process, and ends at one or more undesirable consequences.
2. It is used to represent the possible steps leading to a failure or accident.
3. It uses a series of branches that relate the proper operation and/or failure of a system with the ultimate consequences.
4. It is a quick identification of the various hazards that could result from a single initial event.
5. It is beneficial in examining the possibilities and consequences of a failure.
6. Usually it does not quantify (although it can) the potential of the event occurring.
7. It can be incomplete if all the initial events are not identified.

Thus, the use of event trees is sometimes limited for hazard analysis because they usually do not quantify the potential of the event occurring. They may also be incomplete if all the initial occurrences are not identified. Their use is beneficial in examining, rather than evaluating, the possibilities and consequences of a failure. For this reason, a fault tree analysis should supplement this model to establish the probabilities of the event tree branches. This topic is introduced in Problem FET 3.

FET 3: RUNAWAY CHEMICAL REACTION I

A runaway chemical reaction can lead to a fire, an explosion, or massive pollutant emissions. It occurs if coolers fail (A) or there is a bad chemical batch (B). Coolers fail only if both cooler 1 fails (C) and cooler 2 fails (D). A bad chemical batch occurs if there is a wrong mix (E) or there is a process upset (F). A wrong mix occurs only if there is an operator error (G) and instrument failure (H). Construct a fault tree.

Solution

Begin with the top event for this problem shown in the Figure 21.10. Then generate the first branch of the fault tree, applying the logic gates (see Figure 21.11), as well as the second branch of the fault tree, applying the logic gates as shown in Figure 21.12. Finally, generate the third branch of the fault tree, applying the logic gates as shown in Figure 21.13.

FIGURE 21.10 Top event of fault tree.

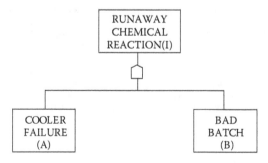

FIGURE 21.11 Fault tree with first branch.

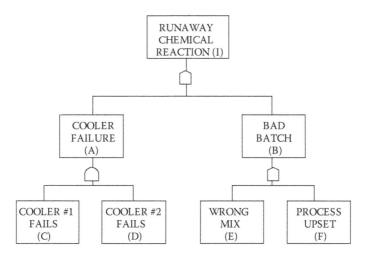

FIGURE 21.12 Fault tree with second branch.

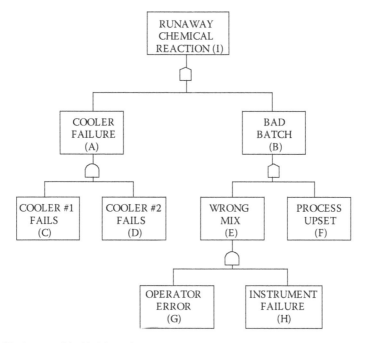

FIGURE 21.13 Fault tree with third branch.

The reader should note the behavior of a fault tree, including the following:

1. A fault tree works backward from an undesirable event or ultimate consequence to the possible causes and failures.
2. It relates the occurrence of an undesired event to one or more preceding events.
3. It "chain-links" basic events to intermediate events that are in turn connected to the top event.
4. It is used in the calculation of the probability of the top event.
5. It is based on the most likely or credible events that lead to a particular failure or accident.
6. Its analysis includes human error as well as equipment failure.

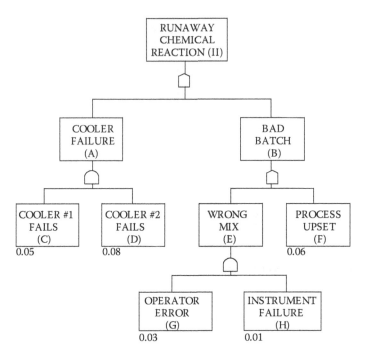

FIGURE 21.14 Runaway chemical reaction probabilities.

FET 4: RUNAWAY CHEMICAL REACTION II

Refer to Problem FET 3. If the following annual probabilities are provided by the plant engineer, calculate the probability of a runaway chemical reaction occurring in a year's time given the following probabilities:

$$P(C) = 0.050$$
$$P(D) = 0.080$$
$$P(F) = 0.060$$
$$P(G) = 0.030$$
$$P(H) = 0.010$$

Solution

Redraw Figure 21.13 in Problem FET 3 and insert the given probabilities (see Figure 21.14). Calculate the probability that the runaway reaction will occur:

$$P = (0.05)(0.08) + (0.01)(0.03) + 0.06$$

$$= 0.0040 + 0.0003 + 0.06$$

$$= 0.064$$

Note that the process upset, F, is the major contributor to the probability.
The reader should also note that if A and B are not mutually exclusive then

$$P = 0.004 + 0.0603 - (0.004)(0.0603)$$

$$= 0.06406$$

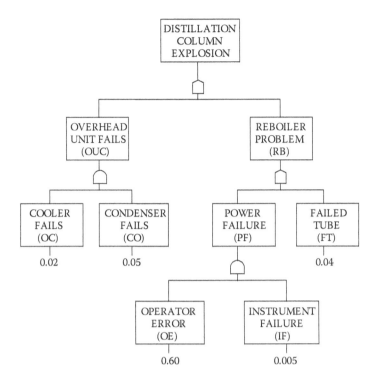

FIGURE 21.15 Distillation column explosion fault tree.

FET 5: DISTILLATION COLUMN FAILURE I

A distillation column explosion can occur if the overhead cooler fails (*OC*) and condenser fails (*CO*) or there is a problem with the reboiler (*RB*). The overhead unit fails (*OUC*) only if both the cooler fails (*OC*) and the condenser fails (*CO*). Reboiler problems develop if there is a power failure (*PF*) or there is a failed tube (*FT*). A power failure occurs only if there are both operator error (*OE*) and instrument failure (*IF*). Construct a fault tree.

Solution

The fault tree is given in Figure 21.15.

FET 6: DISTILLATION COLUMN FAILURE II

Refer to Problem FET 5. If the following annual probabilities are provided by the plant engineer, calculate the probability of an explosion in a year's time.

$$P(OC) = 0.02$$
$$P(CO) = 0.05$$
$$P(FT) = 0.04$$
$$P(OE) = 0.60$$
$$P(IF) = 0.005$$

Solution

Based on the data provided,

Probability and Statistics Applications for Environmental Science

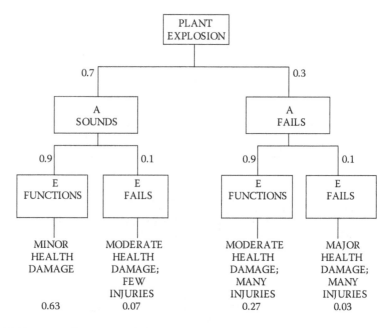

FIGURE 21.16 Event tree for plant explosion.

$$P(\text{explosion}) \approx (0.02)(0.05) + (0.005)(0.6) + 0.04$$

$$\approx 0.001 + 0.003 + 0.04$$

$$\approx 0.0413$$

Note that the major contributor to an explosion is the failed tube (*FT*), whose probability is 0.040.

FET 7: PLANT EXPLOSION

An explosion followed immediately by an emission occurs in a plant, and alarm (A) sounds with probability 0.7. The emission control system (E) functions with probability 0.9 whether or not the alarm sounds. The consequences are minor health damage (A sounds, E works), moderate health damage with few injuries (A sounds, E fails), moderate health damage with many injuries (A fails, E works), and major damage with many injuries (A fails, E fails). Construct an event tree and indicate the probabilities for each of the four consequences.

Solution

This is a "takeoff" on Problem FET 2. The event tree for this application is shown in Figure 21.16.

FET 8: DETONATION AND FLAMMABLE TOXIC GAS RELEASE

A risk assessment is being conducted at a chemical plant to determine the consequences of two incidents, with the initiating event being an unstable chemical. The incidents are as follows:

Incident I: an explosion resulting from detonation of an unstable chemical
Incident II: a release of a flammable toxic gas

Incident I has one possible outcome, the explosion, the consequences of which are assumed to be unaffected by weather conditions. Incident II has several possible outcomes, which, for purposes of simplification, are reduced to the following:

Outcome IIA: vapor-cloud explosion, caused by ignition of the gas released, centered at the release point, and unaffected by weather conditions
Outcome IIB: toxic cloud extending downwind and affected by weather conditions

Again for purposes of simplification, only two weather conditions are envisioned, a northeast wind and a southwest wind. Associated with these two wind directions are events IIB1 and IIB2:

Event IIB1: toxic cloud to the southwest
Event IIB2: toxic cloud to the northeast

Prepare an event tree for incident I and incident II.
(**Note:** This was an actual analysis conducted at a chemical plant.)

Solution

The event tree is provided in Figure 21.17.
Additional details, including numerical data and calculations, can be found in the literature [2–4].

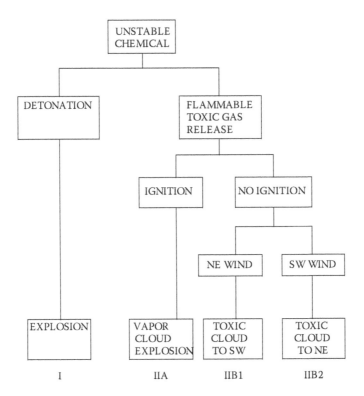

FIGURE 21.17 Event tree showing outcomes of two incidents involving an unstable chemical.

REFERENCES

1. Lees, *Loss Prevention in the Chemical Process Industries*, Butterworth, 1980.
2. Theodore, L., Reynolds, J., and Taylor, F., *Accident and Emergency Management*, John Wiley & Sons, Hoboken, NJ, 1989.
3. Flynn, A.M. and Theodore, L., *Health, Safety and Accident Management in the Chemical Process Industries*, Marcel Dekker, New York, 1992.
4. Theodore, L., Reynolds, J., and Morris, K., *Health, Safety and Accident Prevention: Industrial Applications, a Theodore Tutorial*, East Williston, New York, 1992.

Section III

Contemporary Applications

22 CIM. Confidence Intervals for Means

INTRODUCTION

The sample mean, $\overline{X}$, constitutes a so-called point estimate of the mean, μ, of the population from which the sample was selected at random. Instead of a *point* estimate, an *interval* estimate of μ may be required along with an indication of the confidence that can be associated with the interval estimate. Such an interval estimate is called a *confidence interval*, and the associated confidence is indicated by a *confidence coefficient*. The length of the confidence interval varies directly with the confidence coefficient for fixed values of n, the sample size; the larger the value of n, the shorter the confidence interval. Thus, for fixed values of the confidence coefficient, the limits that contain a parameter with a probability of 95% (or some other stated percentage) are defined as the 95% (or that other percentage) *confidence limits* for the parameter; the interval between the confidence limits is referred to as the *confidence interval*.

For normal distributions, the confidence coefficient Z can be obtained from the standard normal table for various confidence limits or corresponding levels of significance for μ. Some of these values are provided in Table 22.1. Thus, this table can be employed to obtain the probability of a value falling inside or outside the range $\mu \pm Z\sigma$. For example, when $Z = 2.58$, one can say that the level of significance is 1%. The statistical interpretation of this is as follows: if an observation deviates from the mean by at least ± 2.58, σ, the observation is significantly different from the body of data on which the describing normal distribution is based. Further, the probability that this statement is in error is 1%; i.e., the conclusion drawn from a rejected observation will be wrong 1% of the time.

From Table 22.1, one can also state that there is a 99% probability that an observation will fall within the range $\mu \pm 2.58$. This degree of confidence is referred to as the *99% confidence level*. The limits, $\mu - 2.58\sigma$ and $\mu + 2.58\sigma$, are defined as the *confidence limits*, whereas the difference between the two values is defined as the *confidence interval*. Once again, the example in the preceding text essentially states that the actual (or true) mean lies within the interval $\mu - 2.58\sigma$ and $\mu + 2.58\sigma$ with a 99% probability of being correct.

The foregoing analysis can be extended to provide the difference of two population means, i.e., $X_2 - X_1$. For this case, the confidence limits are

$$\overline{X_2} - \overline{X_1} \pm Z\sqrt{\frac{\sigma_2^2}{n_2} + \frac{\sigma_1^2}{n_1}} \tag{22.1}$$

This problem set also serves to introduce Student's t distribution. It is common to use this distribution if the sample size is small. For a random sample of size n selected from a normal population, the term $(\overline{X} - \mu)/(s/\sqrt{n})$ has Student's distribution with $(n - 1)$ degrees of freedom. *Degrees of freedom* is the label used for the parameter appearing in the Student's distribution pdf in Equation 22.2.

TABLE 22.1
Confidence Levels and Levels of Significance

Confidence Level (%)	Level of Significance (%)	Z
80.00	20.00	1.28
90.00	10.00	1.65
95.00	5.00	1.96
95.45	4.55	2.00
98.00	2.00	2.33
99.00	1.00	2.58
99.73	0.27	3.00
99.90	0.10	3.29
99.99	0.01	3.89

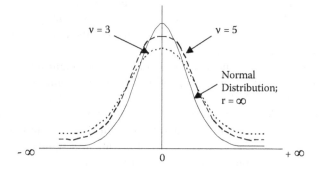

FIGURE 22.1 Student's t distribution.

$$f(t) = \frac{\Gamma\left(\frac{\upsilon+1}{2}\right)}{\sqrt{\pi \upsilon}\,\Gamma\left(\frac{\upsilon}{2}\right)}\left(1+\frac{t^2}{\upsilon}\right)^{-\frac{\upsilon+1}{2}} ; \quad -\infty < t < \infty; \quad \Gamma = \text{Gamma function} \qquad (22.2)$$

The graph of the pdf of Student's distribution is symmetric about 0 for all values of the parameter. The graph is similar to but not as peaked as the standard normal curve. (See Figure 22.1, Problem CIM 5).

Once again, assume X is normally distributed with mean μ and variance σ^2. Let $\overline{X}$ and s^2 represent the corresponding sample estimates on 10 random samples. For this case, Z is still given by

$$Z = \frac{\overline{X}-\mu}{\sigma/\sqrt{10}} \qquad (22.3)$$

with $n = 10$. However, t is given as

$$t = \frac{\overline{X} - \mu}{s/\sqrt{n}} \tag{22.4}$$

where t is drawn from Student's t distribution with $n - 1$ degrees of freedom (usually designated as ν). If $t_{0.01}$ represents the value of t such that the probability is 1% that $|t| > t_{0.01}$, one can conclude that the probability is 99% that

$$\frac{\overline{X} - \mu}{s/\sqrt{10}} < t_{0.01}$$

The preceding equation may also be rewritten as

$$\overline{X} - t_{0.01} \frac{s}{\sqrt{10}} < \mu < \overline{X} + t_{0.01} \frac{s}{\sqrt{10}}$$

For this two-sided test, the appropriate value of t from Table A.2 (Appendix A) is 3.25 (probability of 0.005 for each tail).

$$\overline{X} - \frac{3.25s}{\sqrt{10}} < \mu < \overline{X} + \frac{3.25s}{\sqrt{10}} = 0.99$$

This may also be written as

$$P(-3.25 < T < 3.25) = 0.99 = 99\%$$

Finally, it should be noted that the procedure given in the preceding text provides confidence limit information on both sides of the mean, often referred to as the two *tails* of the interval. For information on one side of the tail (or a "one-tailed" test), the interval is located on one side of the mean, with the level of significance totally associated with that side. Applications include testing whether one mean is better than another, as opposed to whether it is worse.

PROBLEMS AND SOLUTIONS

CIM 1: LARGE SAMPLES: MEANS

Briefly describe the distribution of the sample mean for large populations.

Solution

For large samples (sample size greater than 30) the sample mean, $\overline{X}$, is approximately normally distributed with mean μ and standard deviation $\sigma/\sqrt{n}$, where σ is the population standard deviation. Therefore, for large samples $(\overline{X} - \mu)/(\sigma/\sqrt{n})$ is approximately distributed as a standard normal variable. If σ is unknown, s, the sample standard deviation, can replace it.

CIM 2: NANOCHEMICAL IMPURITY

An environmental scientist has been informed that the standard deviation of an impurity in a nanochemical is 0.23%. The scientist later drew 36 samples from a new batch of the chemical, and the average impurity was 1.92%. Calculate the 99% confidence interval for the true mean. Assume that X is normally distributed.

Solution

The standard normal variable is

$$Z = \frac{\overline{X} - \mu}{\sigma/\sqrt{n}} \tag{22.3}$$

Noting that

$$\overline{X} = 1.92$$

$$\sigma = 0.23$$

$$n = 36$$

one obtains

$$Z = \frac{1.92 - \mu}{0.038}$$

From Table A.1A (Appendix A) one determines that the 99% probability is between −2.57 and + 2.57. Therefore,

$$P(-2.57 < \frac{\overline{X} - \mu}{0.038} < 2.57) = 0.99$$

$$P\,(-0.0985 < \overline{X} - \mu < 0.0985) = 0.99$$

Thus, the probability is approximately 99% that $\overline{X}$ will be within 0.0985 of the true mean μ. With the observed value of $\overline{X}$ = 1.92, one can say there is a 99% confidence that the true mean is in the interval 1.92 ± 0.0985. Thus,

$$1.822 < \mu < 2.019$$

CIM 3: OZONE MEASUREMENT

Ozone measurements in parts per million in an air pollution study were obtained at a sample of 40 sites selected at random in a large city. The sample mean, X, was 10.1. The sample standard deviation, s, was 2.3. Obtain a 95% confidence interval for the population mean ozone measurement for this city.

Solution

First, note that the sample size is large. Select a statistic involving the sample mean and population mean and having a known pdf under the assumptions $(\overline{X}-\mu)/(\sigma/\sqrt{n})$ is approximately distributed as a standard normal variable for $n = 40$. Construct an inequality concerning the statistic selected, such that the probability that the inequality is true equals the desired confidence coefficient. From the normal table, $P(-1.96 < Z < 1.96) = 0.95$ if Z is a standard normal variable. Therefore,

$$P\left(-1.96 < \frac{\overline{X}-\mu}{\sigma/\sqrt{n}} < 1.96\right) = 0.95$$

Solve the preceding equation for the population mean μ,

$$\left(-1.96 < \frac{\overline{X}-\mu}{\sigma/\sqrt{n}} < 1.96\right)$$

$$-1.96(\sigma/\sqrt{n}) < \overline{X}-\mu < 1.96(\sigma/\sqrt{n})$$

$$-\overline{X}-1.96(\sigma/\sqrt{n}) < -\mu < -\overline{X}+1.96(\sigma/\sqrt{n})$$

Multiplication by -1 requires reversal of the inequality signs,

$$\overline{X}+1.96(\sigma/\sqrt{n}) > \mu > \overline{X}-1.96(\sigma/\sqrt{n})$$

or

$$\overline{X}-1.96(\sigma/\sqrt{n}) < \mu < \overline{X}+1.96(\sigma/\sqrt{n})$$

Substitute the observed values from the sample to obtain the 95% confidence interval for μ. Note that $\overline{X} = 10.1$, $n = 40$, σ is unknown and is replaced by s, the sample standard deviation whose value is 2.3.

$$10.1-1.96(2.3/\sqrt{40}) < \mu < 10.1+1.96(2.3/\sqrt{40})$$

$$9.39 < \mu < 10.81$$

Therefore, $9.39 < \mu < 10.81$ is the 95% confidence interval for μ.

CIM 4: PCB Superfund Data

A sample of 82 PCB readings at a Superfund site has a mean concentration of 650 ppm and a standard deviation of 42 ppm. Obtain the 98% confidence limits for the mean concentration.

Solution

Based on the problem statement,

$$\overline{X} = 650; \quad s = 42; \quad n = 82$$

Assume that $\sigma = s$. Therefore,

$$\frac{\sigma}{\sqrt{n}} = \frac{s}{\sqrt{n}} = \frac{42}{\sqrt{82}}$$

$$= 4.638$$

For a two-tailed estimated (with Z normally distributed),

$$P(-2.33 < Z < 2.33) = 0.98$$

or

$$P\left(-2.33 < \frac{\overline{X} - \mu}{\sigma/\sqrt{n}} < 2.33\right) = 0.98$$

Solving gives

$$\mu = 650 \pm 10.79$$

or

$$639.21 < \mu < 660.79$$

CIM 5: VARYING CONFIDENCE INTERVALS FOR MEANS FROM LARGE POPULATIONS

Refer to Problem CIM 3. Obtain the 90 and 99% confidence interval for the mean ozone measurement.

Solution

For the 99% confidence interval, replace the coefficient 1.96 by 2.58 (see Table A.1A of Appendix A). Therefore,

$$10.1 - (2.58)(0.364) < \mu < 10.1 + (2.58)(0.364)$$

$$9.16 < \mu < 11.04$$

For the 90% confidence interval, replace 1.96 by 1.65, so that

$$10.1 - (1.65)(0.364) < \mu < 10.1 + (1.65)(0.364)$$

$$9.50 < \mu < 10.70$$

Note that as the "confidence" decreases, the "interval" increases.

CIM 6: SMALL SAMPLE MEANS

Briefly describe the distribution of means for small samples.

Solution

As indicated in the Introduction, Student's t distribution is employed when dealing with small samples. The graph of the pdf of Student's distribution is symmetric about 0 for all values of the parameter. The graph is similar to but not as peaked as the standard normal curve (see Figure 22.1). The graph approaches the normal distribution in the limit when the degrees of freedom approach infinity. The term t_r has also been used to designate a random variable having Student's distribution with r degrees of freedom. When the sample size is small ($n < 30$), σ is unknown, and the population sampled is assumed to be normal, the statistic $(\overline{X}-\mu)/(s\sqrt{n})$ can, as described earlier, be used to generate confidence intervals for the population mean μ.

CIM 7: CONFIDENCE INTERVAL: T DISTRIBUTION

Discuss the confidence range or interval as it applies to the t distribution.

Solution

As noted in the discussion of the normal distribution, $\overline{X}$ is an estimate of the true mean μ. The t function provides the distribution of deviations of X from μ in terms of probabilities (or relative frequencies). If one rewrites the t equation in terms of plus or minus, the true mean for this class is given by

$$\mu = \overline{X} \pm ts$$

In ordinary language, the true mean can be said, with the tabulated probability of error, to be within the range of the calculated mean included in the limits of plus and minus t times the estimated standard deviation of the mean.

It should be once again noted that one is merely applying the confidence interval numbers to a given experiment. It is obviously incorrect to state that the probability is 0.95 (or 0.99, etc.) that the interval contains μ; the latter probability is either 1 or 0, depending on whether μ does or does not lie in this interval. What it does mean is that μ will lie within the interval 95 out of 100 times. It is only when the random interval $X \pm ts$ is considered that one can make correct probability statements of the type desired.

CIM 8: SMALL SAMPLE CONFIDENCE INTERVAL FOR POPULATION MEAN
GAS MILEAGE

The distance traveled on a gallon of gasoline was measured for a random sample of 16 cars of a particular make and model. The sample mean was 28 mi, and the sample standard deviation was 2.6 mi. Obtain a 95% confidence interval for the population mean gas mileage, μ, under the assumption that the sample comes from a normal population.

Solution

Note that the sample size is small. Select a statistic involving the sample mean and population mean and having a known pdf under the assumption that $(\overline{X}-\mu)/(s/\sqrt{n})$ has a Student's distribution with $(n-1)$ degrees of freedom. Once again, construct an inequality concerning the statistic selected, such that the probability that the inequality is true equals the desired confidence coefficient. From the table of Student's t distribution (Table A.2, Appendix A),

$$P(-2.131 < v_r < 2.131) = 0.95$$

$$P(-2.131 < v_{n-1} < 2.131) = 0.95$$

$$P(-2.131 < v_{15} < 2.131) = 0.95$$

Therefore,

$$P\left(-2.131 < \frac{\overline{X} - \mu}{s/\sqrt{n}} < 2.131\right) = 0.95$$

The preceding equation may be solved for the population mean μ.

$$\left(-2.131 < \frac{\overline{X} - \mu}{s/\sqrt{n}} < 2.131\right)$$

$$-2.131(s/\sqrt{n}) < \overline{X} - \mu < 2.131(s/\sqrt{n})$$

$$-\overline{X} - 2.131(s/\sqrt{n}) < -\mu < -\overline{X} + 2.131(s/\sqrt{n})$$

Multiplication by -1 requires reversal of the inequality signs,

$$\overline{X} + 2.131(s/\sqrt{n}) > \mu > \overline{X} - 2.131(s/\sqrt{n})$$

$$\overline{X} - 2.131(s/\sqrt{n}) < \mu < \overline{X} + 2.131(s/\sqrt{n})$$

Select a basis of 1 gal. Substitute the observed values from the sample to obtain the 95% confidence interval for μ. Note that $\overline{X} = 28$, $n = 16$, and $s = 2.6$.

$$28 - 2.131(2.6/\sqrt{16}) < \mu < 28 + 2.131(2.6/\sqrt{16})$$

$$26.61 < \mu < 29.39$$

Therefore, $26.61 < \mu < 29.39$ is the 95% confidence interval for μ.

CIM 9: SINGLE-TAIL APPLICATION OF STUDENT'S DISTRIBUTION

Obtain the confidence coefficient t for either a left or right tail of Student's distribution in the following cases:

1. For $v = 10$, tail area $= 0.05$
2. For $v = 10$, tail area $= 0.01$
3. For $v = 20$, tail area $= 0.10$
4. For $v = 20$, tail area $= 0.025$

5. For $v = 100$, tail area $= 0.05$
6. For $v = \infty$, tail area $= 0.05$

Solution

Refer to Table A.2 (Appendix A):

1. With $v = 10$, 95% (0.95) tail, $t_{10} = 1.81$
2. With $v = 10$, 99% (0.99) tail, $t_{10} = 2.76$
3. With $v = 20$, 90% (0.90) tail, $t_{20} = 1.32$
4. With $v = 20$, 97.5% (0.975) tail, $t_{20} = 2.09$
5. With $v = 100$, 95% (0.95) tail, t_{100} (interpolated) $= 1.66$
6. With $v = \infty$, 95% (0.95) tail, $t_{\infty} = 1.645$ (normal distribution)

Note that the last calculated value corresponds exactly to that provided by a normal distribution, i.e., 1.645; see Table A.1A (Appendix A).

CIM 10: Varying Confidence Intervals for Means from a Small Population

Refer to Problem CIM 8. Obtain the 98 and 99% confidence interval for the mean gas mileage.

Solution

For the 98% confidence interval, replace the coefficient 2.13 by 2.60 (see Table A.2, Appendix A). Therefore,

$$28 - (2.60)(0.65) < \mu < 28 + (2.60)(0.65)$$

$$26.31 < \mu < 29.69$$

For the 99% confidence interval, replace 2.13 by 2.95, so that

$$28 - (2.95)(0.65) < \mu < 28 + (2.95)(0.65)$$

$$26.08 < \mu < 29.92$$

23 CIP. Confidence Intervals for Proportions

INTRODUCTION

If a random variable X has a binomial distribution, then for large n the random variable X/n is approximately normally distributed with mean p and standard deviation equal to the square root of pQ/n. If n is the size of a random sample from a population in which p is the proportion of elements classified as "success" because of the possession of a specified characteristic, then the large sample distribution of X/n, the sample proportion having the specified characteristic, serves as the basis for constructing a confidence interval for p, the corresponding population proportion. By "n is large" it is meant that np and $n(1 - p)$ are both greater than 5.

One can generate an inequality whose probability equals the desired confidence coefficient using the large sample distribution of the sample proportion. Note, once again, that X/n, the sample proportion, is approximately normally distributed with mean p and standard deviation equal to the square root of pQ/n. Therefore, $(X/n - p)/(pQ/n)^{1/2}$ is approximately a standard normal variable and

$$P\left(-Z < \frac{X/n - p}{\sqrt{pQ/n}} < Z\right) = \text{confidence interval, fractional basis} \qquad (23.1)$$

where Z is obtained from the table of the standard normal distribution as the value such that the area over the interval from $-Z$ to Z is the stated fractional basis. This inequality can then be converted into a statement about P:

$$X/n - Z\sqrt{pQ/n} < P < X/n + Z\sqrt{pQ/n} \qquad (23.2)$$

The endpoints of the statement may then be evaluated after replacing p and Q in the endpoints by the observed value of X/n and $1 - X/n$, respectively.

The large sample distribution of the sample proportion X/n also provides the basis for determining the sample size n for estimating the population proportion p with maximum allowable error E and specified confidence. This is illustrated in the last three problems in this chapter.

PROBLEMS AND SOLUTIONS

CIP 1: Large Sample Confidence Interval for Proportion of Accidents Due to Unsafe Working Conditions I

In a random sample of 300 accidents reported in a chemical industry, 173 were found to be due to unsafe working conditions. Obtain a 99% confidence interval for the proportion of all reported accidents in this industry that are due to unsafe working conditions.

Solution

For this problem, the population is the set of all accidents reported in a chemical industry, and the population proportion to be estimated is the proportion of accidents that are due to unsafe working conditions. The corresponding sample proportion is the proportion of the random sample of 300 reported accidents that were found to be due to unsafe working conditions, i.e., 173/300, the observed value of X/n.

The term, X/n, the sample proportion, may be assumed to be approximately normally distributed with mean p and standard deviation equal to the square root of pQ/n. Therefore, $(X/n-p)/(pQ/n)^{1/2}$ is approximately a standard normal variable. From Equation 23.1

$$P\left(-2.58 < \frac{X/n-p}{\sqrt{pQ/n}} < 2.58\right) = 0.99$$

where 2.58 is obtained from the table of the standard normal distribution as the value such that the area over the interval from –2.58 to 2.58 is 0.99. This equation may be rewritten in terms of P:

$$X/n - 2.58\sqrt{pQ/n} < P < X/n + 2.58\sqrt{pQ/n}$$

Replacing P in the endpoints by the observed value of X/n, and Q by $1 - X/n$ gives

$$\frac{173}{300} - 2.58\sqrt{\frac{\left(\frac{173}{300}\right)\left(1-\frac{173}{300}\right)}{300}} < P < \frac{173}{300} + 2.58\sqrt{\frac{\left(\frac{173}{300}\right)\left(1-\frac{173}{300}\right)}{300}}$$

$$0.5767 - (2.58)(0.0285) < P < 0.5767 + (2.58)(0.0285)$$

$$0.5767 - 0.0735 < P < 0.5767 + 0.0735$$

$$0.503 < P < 0.650$$

Thus, $0.503 < P < 0.650$ is the 99% confidence interval for the population proportion of accidents that are due to unsafe working conditions.

CIP 2: LARGE SAMPLE CONFIDENCE INTERVAL FOR PROPORTION OF ACCIDENTS DUE TO UNSAFE CONDITIONS II

Refer to Problem CIP 1. Obtain a 90% confidence interval for the proportion of all reported accidents. Repeat the calculation for a 95% confidence interval.

Solution

Referring to the solution for Problem CIP 1, replace the 2.58 term (for 99%) by 1.645 (for 90%):

$$P\left(-1.645 < \frac{(X/n)-p}{\sqrt{pQ/n}} < 1.645\right) = 0.90$$

Solving for P gives

$$0.5767 - (1.645)(0.0285) < P < 0.5767 + (1.645)(0.0285)$$

$$0.5296 < P < 0.6234$$

For the 95% confidence interval, replace 2.58 by 1.960. Thus,

$$P\left(-1.960 < \frac{X/n - p}{\sqrt{pQ/n}} < 1.960\right) = 0.95$$

Solving for P gives

$$0.5767 - (1.960)(0.0285) < P < 0.5767 + (1.960)(0.0285)$$

$$0.5206 < P < 0.6324$$

As the confidence interval percentage increases, the interval increases.

CIP 3: LARGE SAMPLE CONFIDENCE INTERVAL FOR PROPORTION OF ACCIDENTS DUE TO UNSAFE WORKING CONDITIONS III

Refer to Problem CIP 1. Repeat the calculation if 284 accidents were reported the following year and 138 of these were found to be due to unsafe working conditions.

Solution

For this problem, the sample proportion is

$$\frac{X}{n} = \frac{138}{284}$$

$$= 0.486$$

Therefore,

$$0.486 - 2.58\sqrt{\frac{(0.486)(1-0.486)}{284}} < P < 0.486 + 2.58\sqrt{\frac{(0.486)(1-0.486)}{284}}$$

$$0.486 - (2.58)(0.0297) < P < 0.486 + (2.58)(0.0297)$$

$$0.486 - 0.0765 < P < 0.486 + 0.0765$$

$$0.410 < P < 0.563$$

Thus, $0.4106 < P < 0.563$ is the 99% confidence interval for the population proportion of accidents that are due to unsafe working conditions.

CIP 4: Large Sample Confidence Interval for Proportion of Accidents Due to Unsafe Working Conditions IV

Refer to Problem CIP 3. Resolve the problem for the following confidence intervals.

1. 95%
2. 99.5%
3. 99.9%

Solution

In this problem, the describing equation becomes (where $Z_{\alpha/2}$ is now exmployed to note the appropriate Z value for a two-sided test).

$$0.486 - (Z_{\alpha/2})(0.0297) < P < 0.486 + (Z_{\alpha/2})(0.0297)$$

1. For 95%, $Z_{\alpha/2} = 1.96$ so that

$$0.486 - (1.96)(0.0297) < P < 0.486 + (1.96)(0.0297)$$

$$0.428 < P < 0.544$$

2. For 99.5%, $Z_{\alpha/2} = 2.81$, and

$$0.486 - (2.81)(0.0297) < P < 0.486 + (2.81)(0.0297)$$

$$0.403 < P < 0.570$$

3. For 99.9%, $Z_{\alpha/2} = 3.30$, and

$$0.486 - (3.30)(0.0297) < P < 0.486 + (3.30)(0.0297)$$

$$0.388 < P < 0.584$$

Once again, the reader should note that as the "confidence percentage" increases, the confidence interval correspondingly increases.

CIP 5: Defective Thermometers

One percent of the thermometers produced by an instrumentation manufacturer are usually defective. Generate a 95% confidence interval for the proportion of defective thermometers in a 1000-lot batch.

Solution

For this case,

$$p = 0.01$$

$$Q = 0.99$$

$$\sqrt{pQ/n} = \sqrt{(0.01)(0.99)/(1000)} = 0.00315$$

Thus, with $Z_{0.025} = 1.96$ for this two-tailed test,

$$0.01 - (1.96)(0.00315) < P < 0.01 + (1.96)(0.00315)$$

$$0.00383 < P < 0.0162$$

Because the final answer must be rounded off to whole numbers, one can conclude (approximately) that for a 95% confidence interval the number of thermometers in a 1000-lot batch is in a range of 4 to 16.

CIP 6: SAMPLE SIZE FOR ESTIMATING DEFECTIVE PROPORTION

Outline how to calculate the sample size for estimating the defective population proportion, p, with maximum allowable error, E, and specified confidence.

Solution

Suppose the specified confidence is 95%. Then the sample size n must be such that

$$P(-E < X/n - p < E) = 0.95$$

Solving for P gives

$$P\left(-\frac{E}{\sqrt{\dfrac{pQ}{n}}} < \frac{\dfrac{X}{n} - p}{\sqrt{\dfrac{pQ}{n}}} < \frac{E}{\sqrt{\dfrac{pQ}{n}}}\right) = 0.95$$

As before, for large n,

$$\frac{\dfrac{X}{n} - p}{\sqrt{\dfrac{pQ}{n}}}$$

is approximately distributed as a standard normal variable. From the table of the distribution of a standard normal variable, it can be verified that the variable in question lies between -1.96 and 1.96 (two-sided) with probability 0.95. Setting

$$\frac{E}{\sqrt{\dfrac{pQ}{n}}} = 1.96 = Z_{0.025}$$

and solving for n gives

$$n = \frac{pQ(1.96)^2}{E^2} = \frac{p(1-p)(1.96)^2}{E^2}$$

as the sample size required to estimate the population proportion defective, p, with maximum allowable error E and 95% confidence.

In general, if the required confidence is $1 - \alpha$, then the formula for the sample size n required becomes

$$n = \frac{pQZ_{\alpha/2}^2}{E^2}$$

where $Z_{\alpha/2}$ is obtained from a table of the standard normal distribution as the value such that a standard normal variable lies between $-Z_{\alpha/2}$ and $Z_{\alpha/2}$ with probability $1 - \alpha$.

CIP 7: SAMPLE SIZE FOR DEFECTIVE TRANSISTORS

How large a random sample is required to estimate the proportion, p, of defective transistors in a large lot submitted for inspection, if the maximum allowable error of the transistor variable is 0.02 (consistent units) and the confidence required is 90%? Determine the sample size when it is known that P is at most 10%.

Solution

The maximum allowable error, E, and the required confidence, $1 - \alpha$, are

$$E = 0.02$$

$$1 - \alpha = 0.90$$

Determine $Z_{\alpha/2}$ from $P(-Z_{\alpha/2} < Z < Z_{\alpha/2}) = 1$ where Z is a standard normal variable.

$$P(-Z_{0.05} < Z < Z_{0.05}) = 0.90$$

From the table of the standard normal distribution, $Z_{0.05} = 1.65$. If an estimate of maximum p is known, substitute this for p in the formula,

$$n = \frac{(0.10)(0.90)1.65^2}{0.02^2} = 613$$

The minimum sample size required is therefore 613.

CIP 8: SAMPLE SIZE CALCULATION WHEN P IS UNKNOWN

Refer to Problem CIP 7. Determine the minimum sample size required when p is unknown.

Solution

Because nothing is known about p, it may be assigned a value of 0.5. (Note that the term $p(1-p)$ has a maximum when $p = 0.5$). Therefore,

$$n = \frac{(0.5)(0.5)1.65^2}{0.02^2} = 1702$$

The minimum sample size required is therefore 1702, because $p(1-p)$ has been set at its maximum value.

24 HYT. Hypothesis Testing

INTRODUCTION

In hypothesis testing, one makes a statement (hypothesis) about an unknown population parameter and then uses statistical methods to determine (test) whether the observed sample data support that statement. Note that a statistical hypothesis is an assumption made about some parameter, i.e., about a statistical measure of a population. One may also define hypothesis as a conjectural statement about one or more parameters. Synonyms for hypothesis could include assumption, or guess. A hypothesis can be tested to verify its credibility.

An important notion to understand is the relationship of the null hypothesis to the alternative hypothesis (or hypotheses). In all applications, the null hypothesis and alternatives are written in terms of population parameters. For example, if μ represents the population mean (temperature, for example), one could write

$$\text{null hypothesis is } H_0: \mu = 100$$

$$\text{alternative hypothesis is } H_1: \mu \neq 100$$

Because statistics are only estimates of a "true" population parameter based on a sample of observations, it is reasonable to expect that the value of the estimate will deviate from the value of the "true" population parameter. One uses a method called hypothesis testing to decide whether the value of the statistic is consistent with a hypothesized value for the population parameter. When an estimate is made, a hypothesis is made concerning an assumed population parameter.

In hypothesis testing, the statistician uses various techniques to decide whether to accept or reject hypotheses. If, for example, someone assumed a population mean of 70 (a thermometer reading or test score, for example), and a sample from that population was selected that had a mean of 80, one might want to perform a test or calculation to see if the population mean of 70 could still be assumed. The statistician would use a statistical technique to either accept or reject the hypothesis that the population mean equals 70.

In the preceding example, a sample mean equal to 80 is being used to test a hypothesized population mean of 70. The notation employed is as follows:

$$\overline{X} = 80 \quad \text{(sample mean)}$$

and the hypothesis is

$$\mu = 70 \text{ (population mean)}$$

As indicated earlier, each null hypothesis is accompanied by an alternative hypothesis. The hypothesis being tested in this example can be written:

$$\mu = 70$$

This hypothesis is defined as the *null hypothesis*, with "null" meaning there is no difference between the hypothesized value (70) and the true value (μ). The null hypothesis is then written as

$$H_0: \mu = 70$$

If the null hypothesis is rejected, an alternative hypothesis is accepted.

It is the null hypothesis that is to be accepted or rejected. In rejecting a null hypothesis, such as $H_0: \mu = 70$ (described earlier), one is accepting another hypothesis (in this case, $\mu \neq 70$), with this second hypothesis referred to as the *alternative hypothesis*. This alternative hypothesis is usually written as

$$H_1: \mu \neq 70$$

For any null hypothesis, one of several alternative hypotheses could be chosen. For example, if

$$H_0: \mu = 90$$

alternatives include

$$H_1: \mu \neq 90$$

or

$$H_1: \mu > 90 \ (\mu \text{ is larger than } 90)$$

or

$$H_1: \mu < 90 \ (\mu \text{ is smaller than } 90)$$

These alternatives will be reviewed again in Chapter 25. Finally, note that it is important when testing hypotheses to know both the null and alternative hypotheses because they will determine which type of statistical test is used.

Details on two other hypothesis tests are provided in the last three problems. Additional details can be found in Chapter 25.

PROBLEMS AND SOLUTIONS

HYT 1: Concept of Hypothesis Testing

Briefly describe the concept of hypothesis testing.

Solution

The concept of hypothesis testing is basic to the realm of statistics, for it is through the acceptance or rejection of the null hypothesis that the statistician can infer certain properties about the population from which a sample is taken. Because the null hypothesis is an assumption made about a population, it is the statistician's job to determine if that assumption should be retained or accepted based on a statistic computed from a sample of that population.

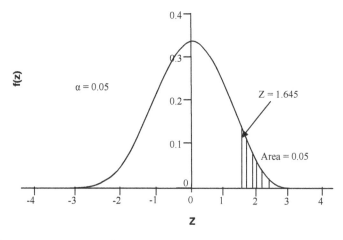

FIGURE 24.1 Critical region for a one-sided test.

HYT 2: Reactor Temperature Test

The mean temperature in a batch reactor is assumed to be 200°C. If a sample of temperature readings from the reactor had a mean of 190°C, would it be reasonable to assume that the mean temperature in the reactor is still 200°C? Set up the null and alternative hypotheses for this scenario.

Solution

For this application,

$$H_0: \mu = 200°C$$

serves as the null hypothesis whereas

$$H_1: \mu \neq 200°C$$

is the alternative hypothesis.

HYT 3: Significance Level

Describe the relationship between a hypothesis and the significance level.

Solution

In testing a statistical hypothesis, the hypothesis is considered to be true if the calculated probability exceeds a given value, α, defined as the significance level. It is considered false if the calculated probability is less than α. Furthermore, if the calculated probability is less than α, (indicating that the hypothesis is false) the result is said to be significant.

If only one "tail" of a curve (distribution) is used in testing the probability associated with a statistical hypothesis, it is termed a one-tailed test (see Figure 24.1). If both tails are employed in the calculation, it is defined as two-tailed (see Figure 24.2). The critical region is described in Problem HYT 4.

HYT 4: Hypothesis Acceptance

Discuss the two options available to accept a hypothesis.

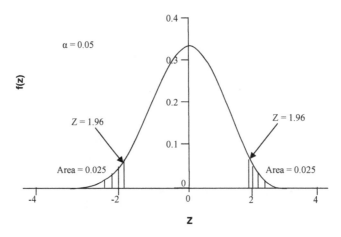

FIGURE 24.2 Critical region for a two-sided test.

Solution

To make a test of the hypothesis, sample data are used to calculate a test statistic. Depending on the definition and value of the test statistic, the primary hypothesis H_0 is either accepted or rejected. The test statistic is determined by the specific probability distribution and by the parameter selected for testing.

The critical region for a one-sided "tail" test appears as the light nonshaded area in Figure 24.1 (see Problem HYT 3) for a significance level of 0.05. As indicated, the variable is assumed to be normally distributed. Figure 24.2 (Problem HYT 3) portrays the critical region (again in the light nonshaded area) for a two-sided "tail" test, i.e., in the area between the two tails.

One may accept the primary hypothesis by either of two ways:

1. The value of the primary statistic (in Figure 24.1 it is Z) must lie in the critical region (or light nonshaded area).
2. The value of the calculated probability exceeds the specified or assigned significance level; i.e., the calculated value of Z produced a dark-shaded area in excess of α. (See Problem HYT 3 for additional details.)

Alternately, the proposal is rejected in the same manner.

HYT 5: PROCEDURE FOR TESTING A HYPOTHESIS

Outline a general procedure that may be employed for testing a hypothesis.

Solution

The procedure for testing a hypothesis is as follows:

1. Choose a probability distribution and a random variable associated with it. This choice may be based on previous experience, intuition, or the literature.
2. Set H_0 and H_1. These must be carefully formulated to permit a meaningful conclusion.
3. Specify the test statistic and choose a level of significance α for the test.
4. Determine the distribution of the test statistic and the "critical region" for the test statistic.
5. Calculate the value of the test statistic from sample data.

Accept or reject H_0 by comparing the calculated value of the test statistic with the critical region.

HYT 6: Two-Sided Test

A single observation value of 16 of a variable from a population believed to be normally distributed has a variance of 9 and a mean of 10. Is it reasonable to conclude, with a level of significance of 0.05, that the observation is from the population?

Solution

Apply the procedure outlined in Problem HYT 5.

1. The probability model is a normal distribution and is specified with $\mu = 10$ and $\sigma^2 = 9$. The variable is the value of X.
2. The null hypothesis is H_0: $X = 10$ and is from a population that is normally distributed with $\mu = 10$ and $\sigma^2 = 9$.
   The alternative hypothesis is H_1: $X \neq 10$ and is not from the population.
3. Because there is only one observation for X, the aforementioned test statistic is simply its standardized value, $Z = (X - \mu)/\sigma$. The standard normal tables give the distribution of this statistic. The level of significance is set with $\alpha = 0.05$.
4. The test observation is distributed normally with $\mu = 10$ and $\sigma^2 = 9$ if H_0 is true. A value of X that is too far above or below the mean should be rejected. A critical region at each end of the normal distribution needs to be determined for $\alpha = 0.05$. As illustrated earlier, 0.025 of the total area under the curve is cut off at each end for a two-tailed test. From normal tables, the limits corresponding to these "tail" areas are $Z = -1.96$ and $Z = 1.96$. Thus, if the single observation falls between these "critical" values, accept H_0.
5. The value for $Z = (16 - 10)/3 = 2.0$.
6. Because $2 > 1.96$, reject H_0 at $\alpha = 0.05$.

HYT 7: Adjusted Level of Significance

Refer to Problem HYT 6. Recalculate the limits for Z if the level of significance had been selected as $\alpha = 0.01$. The corresponding limits for Z would now be

Solution

$$-2.58 < Z < 2.58$$

Because $2.0 < 2.58$, H_0 would be accepted for $\alpha = 0.01$.

HYT 8: Types of Errors

Briefly describe the two types of errors encountered in hypothesis testing.

Solution

Because the sample characteristics may by chance be significantly different from those of the population, it is possible to make an error in accepting or rejecting a stated hypothesis. Two types of error may be defined, together with their probability of occurrence.

Type I: rejecting H_0 when it is true. If α is the probability of rejecting H_0 when it is true, it is referred to as the *level of significance* of the test.

Type II: accepting H_0 when it is false (i.e., when H_1 is true). Here, β is the probability of accepting H_0 when it is false.

Ideally, one would prefer a test that minimizes both types of errors. Unfortunately, as α decreases, β tends to increase, and *vice versa*.

To summarize, the preceding discussion may be restated as follows:

1. If the hypothesis being tested is actually true, and if one concludes from the sample that it is false, a Type I error has been committed.
2. If the hypothesis being tested is actually false, and if one concludes from the sample that it is true, a Type II error has been committed.

One may also summarize the preceding statements in the following manner:

1. A Type I error rejects the null hypothesis, H_0, when it is true.
2. A Type II error fails to reject H_0 when it is false.

HYT 9: CALCULATION OF TYPE I AND TYPE II ERRORS IN TESTING HYPOTHESES

Given a test of a statistical hypothesis, determine α, the probability of a Type I error, and β, the probability of a Type II error, for the following application. Set θ as the unknown fraction of defectives in a box of 20 transistors. The hypothesis H_0: $\theta = 0.1$ is rejected when a random sample of two transistors contains exactly two defectives. The hypothesis H_0 fails to be rejected when the number of defectives is less than 2. Determine α. Also determine β when θ is 0.8.

Solution

Recall that a Type I error rejects the null hypothesis, H_0, when it is true; a Type II error fails to reject H_0 when it is false. Assuming H_0 is true and $\theta = 0.1$, there are 20(0.1) or 2 defectives and 18 nondefectives in the box. Because α is the probability of a Type I error, it represents the probability that both transistors in the sample are defective when there are 2 defectives and 18 nondefectives in the box. By the multiplication theorem, α is the probability that the first transistor drawn is defective times the conditional probability that the second is defective given that the first is defective. Therefore,

$$\alpha = (2/20)(1/19) = 1/190$$

$$= 0.0053$$

If θ is 0.8, there are 20(0.8) or 16 defectives and 4 nondefectives in the box. Based on the information provided, H_0 fails to be rejected when the number of defectives in the sample is either 0 or 1. Proceed to compute β when θ is 0.8, with set β as the probability of a Type II error, i.e., failing to reject H_0 when it is false. When θ is 0.8, H_0 is false, and there are 20(0.8) = 16 defectives in the box. Apply the multiplication theorem to obtain β.

$$P(0 \text{ defective}) = P(\text{1st is nondefective and 2nd is nondefective})$$

Because the multiplication theorem applies,

$$P(0 \text{ defective}) = P(\text{1st is nondefective}) \ P(\text{2nd is nondefective})$$

$$= (4/20)(3/19)$$

$$= 3/95$$

$$= 0.0316$$

$$P(1 \text{ defective}) = P(\text{1st is defective and 2nd is nondefective}) \text{ or}$$
$$(\text{1st is nondefective and 2nd is defective})$$

Because the addition theorem applies,

$$P(1 \text{ defective}) = P(\text{1st is defective and 2nd is nondefective}) +$$
$$P(\text{1st is nondefective and the 2nd is defective})$$

$$= (16/20)(4/19) + (4/20)(16/19)$$

$$= 32/95$$

$$= 0.3368$$

The probabilities of the two ways of getting 1 defective are added in accordance with the addition theorem for mutually exclusive events. Therefore,

$$\beta = P(0 \text{ defectives}) + P(1 \text{ defective})$$

$$= 3/95 + 32/95 = 35/95$$

$$= 7/19$$

$$= 0.3684$$

Note also that the power of a test may be defined as the probability of rejecting H_0 when it is false. Thus, for a stated alternative hypothesis, one may write

$$\text{Power} = 1 - \beta \text{ (fractional basis)}$$

HYT 10: HYPOTHESIS TEST BETWEEN THE MEAN AND ZERO

Outline the hypothesis test between a mean μ and zero.

Solution

The null hypothesis for this case is

$$H_0: \mu = 0$$

Data are often obtained in pairs to compare two properties, two processes, two different materials, etc. The purpose of the test is to determine whether there is a significant difference between the two sets of data in terms of the quantity or measurement involved, or whether the mean of the difference is significantly different from zero. This is a special case of the testing described in the Introduction as to whether a mean is significantly different from some specified value. However, in comparing paired data, the pairs do not have to be measures of the same thing, although the individual measurements in a pair are made in the same conditions. It is the difference within pairs and not the difference between the pairs that is to be tested.

If two different baghouse fabrics are tested for bag life under actual operating conditions, the bag life can vary during the test as different baghouses and operating conditions are encountered. This variation need not be reflected in a variation of difference in bag life between the two fabrics under similar conditions. The purpose of this hypothesis is to determine whether the difference in life between the two fabrics was significant when the analysis is based on the amount of variation in this difference.

HYT 11: Hypothesis Test of the Difference between Two Means

Outline the hypothesis test for the difference between two means.

Solution

The null hypothesis for this case is

$$H_0: \mu_1 = \mu$$

Its purpose is to test whether the means of two different samples may have come from the same population or from populations with the same means. For example, the efficiency of an electrostatic precipitator (ESP) employing one type of wire arrangement produced a collection efficiency is 99.45% over a series of tests. For another ESP or wire arrangement, the mean collection efficiency is 99.63% during a similar test. Is the first unit significantly better than the second? Note that to apply the t test to the difference between the two means, it is necessary to have an estimate of the standard deviation of the measurements.

The appropriate equation to test the difference between two means is as follows:

$$t = \frac{\left| \overline{X}_1 - \overline{X}_2 \right|}{s_p \sqrt{1/n_1 + 1/n_2}} \qquad (24.1)$$

where $\overline{X}_1$ and $\overline{X}_2$ are two individual means, n_1 and n_2 are the number of samples of which $\overline{X}_1$ and X_2 are, respectively, the means, and s_p is the pooled estimate of standard deviation from both sets of data.

The pooled estimate of the standard deviation can be calculated from the following equation:

$$s_p = \sqrt{\frac{\sum (X_1 - \overline{X}_1)^2 + \sum (X_2 - \overline{X}_2)^2}{n_1 + n_2 - 2}} \qquad (24.2)$$

The term t is calculated from Equation 24.1 and is compared with the tabulated values employing the degrees of freedom equal to $n_1 + n_2 - 2$. If the calculated t is larger than the tabulated t at the preselected probability or significance level, then one can conclude that the null hypothesis is false, i.e., the population mean estimated by X_1 is significantly different from the population mean estimated by X_2, with an α chance of being wrong.

HYT 12: Summarizing Results

Summarize hypothesis test details for

1. Estimation of the true mean (Problem HYT 10)
2. The comparison of two means (Problem HYT 11)

Solution

1. The answers for the true mean are detailed in Table 24.1.
2. The answer for the comparison of two means can be found in Table 24.2.

Note

$$\bar{s} = s(X)\sqrt{\frac{1}{n_1} + \frac{1}{n_2}}$$

TABLE 24.1
Estimating the True Mean

	One-Sided	Two-Sided
Test	$\mu > A$	$\mu \neq A$
Hypothesis, H_0	$\mu = A$	$\mu = A$
t test	$\left\|\bar{X} - A\right\| / s$	$\left\|\bar{X} - A\right\| / s$
Degrees of freedom, ν	$n - 1$	$n - 1$
Type I error	False rejection of lesser or equal μ	False rejection of equal μ
Probability of error	$\alpha/2$	$\alpha/2$

TABLE 24.2
Comparison of Two Means

	One-Sided	Two-Sided
Test	$\mu_1 > \mu_2$	$\mu_1 \neq \mu_2$
Hypothesis, H_0	$\mu_1 \leq \mu_2$	$\mu_1 = \mu_2$
t test	$(\bar{X}_1 - \bar{X}_2)/\bar{s}$	$\left\|\bar{X}_1 - \bar{X}_2\right\| / \bar{s}$
Degrees of freedom, ν	$n_1 + n_2 - 2$	$n_1 + n_2 - 2$
Type I error	False rejection of μ, equal or less	False rejection of equal means
Probability of error	$\alpha/2$	α

25 TMP. Hypothesis Test for Means and Proportions

INTRODUCTION

As described in Chapter 24, testing a statistical hypothesis concerning a population mean involves construction of a rule for deciding, on the basis of a random sample from the population, whether or not to reject the hypothesis being tested, i.e., the so-called *null hypothesis*. The test is formulated in terms of a test statistic, i.e., some function of the sample observations. In the case of a large sample ($n > 30$), testing a hypothesis concerning the population mean, μ, involves use of the following test statistic:

$$Z = \frac{\overline{X} - \mu_0}{\dfrac{\sigma}{\sqrt{n}}} \tag{25.1}$$

where $\overline{X}$ is the sample mean, μ_0 is the value of μ specified by H_0, the null hypothesis, σ is the population standard deviation, and n is the sample size. When σ is unknown s, the sample standard deviation, may be employed to estimate it. This test statistic is approximately distributed as a standard normal variable for $n > 30$. Regardless of the sample size, the test statistic is exactly distributed as a standard normal variable when the sample comes from a normal population whose standard deviation, σ, is known.

The values of the test statistic for which the null hypothesis, H_0, is rejected constitute the critical region. The critical region depends on the alternative hypothesis, H_1, and the tolerated region for three alternative hypotheses (see Table 25.1). The term Z_α is the value, which a standard normal variable exceeds with probability α, and $Z_{\alpha/2}$ is defined similarly. Typical values for the tolerated probability of a Type I error are 0.05 and 0.01, with 0.05 the more common selection. These are summarized in Table 25.1.

When the sample is small ($n < 30$), the population standard deviation is unknown, and the sample can be assumed to have come from a normal population, tests of hypotheses concerning the population mean, μ, utilize the following test statistic:

$$t = \frac{\overline{X} - \mu_0}{\dfrac{s}{\sqrt{n}}} \tag{25.2}$$

where $\overline{X}$ is the sample mean, s is the sample standard deviation, n is the sample size, and μ_0 is the value of μ specified by H_0, the null hypothesis. When H_0: $\mu = \mu_0$ is true the test statistic (as described earlier) has Student's t distribution with $n - 1$ degrees of freedom. The critical region, values of the test statistic for which H_0 is rejected, depends on the alternative hypothesis, H_1, and α, the

---

TABLE 25.1
Critical Regions: Normal Distribution

Alternative Hypothesis	Critical Region
$\mu < \mu_0$ (one tail)	$Z < -Z_\alpha$
$\mu > \mu_0$ (one tail)	$Z > Z_\alpha$
$\mu \neq \mu_0$ (two tail)	$Z < Z_{\alpha/2}$ or $Z > Z_{\alpha/2}$

TABLE 25.2
Critical Regions: Student's Distribution

Alternative Hypothesis	Critical Region
$\mu < \mu_0$ (one tail)	$t < -t_\alpha$
$\mu > \mu_0$ (one tail)	$t > t_\alpha$
$\mu \neq \mu_0$ (two tail)	$t < -t_{\alpha/2}$ or $t > t_{\alpha/2}$

---

tolerated probability of Type I error. Listed in Table 25.2 are critical regions for three different alternative hypotheses. The term t_α is the value exceeded with probability by a random variable having Student's distribution with $n - 1$ degrees of freedom, with $t_{\alpha/2}$ defined similarly.

The foregoing analysis can be extended to provide information on the difference (if any) between two means. The following hypothesis is employed to test whether two means are equal. (See Chapter 24 for additional details.)

$$H_0\!: \mu_1 = \mu_2$$

$$H_1\!: \mu_1 < \mu_2 \quad \text{or} \quad \mu_1 > \mu_2 \tag{25.3}$$

The test statistic for this analysis is

$$Z = \frac{\overline{X}_1 - \overline{X}_2 - (\mu_1 - \mu_2)}{\sigma\sqrt{\dfrac{1}{n_1} + \dfrac{1}{n_2}}} \tag{25.4}$$

This equation applies if the distribution of Z is normal, or for large sample sizes. However, if the two populations have different variances, the test statistic becomes

$$Z = \frac{\overline{X}_1 - \overline{X}_2 - (\mu_1 - \mu_2)}{\sqrt{\dfrac{(\sigma_1)^2}{n_1} + \dfrac{(\sigma_2)^2}{n_2}}} \tag{25.5}$$

where $(\sigma_1)^2$ and $(\sigma_2)^2$ are the variances of the two populations.

To compare two means with population variances unknown, the sample variances may be assumed either equal or unequal. If the two unknown variances are assumed to be equal, a *pooled* sample is calculated. (See also Equation 24.3.)

$$(s_p)^2 = \frac{(n_1 - 1)(s_1)^2 + (n_2 - 1)(s_2)^2}{n_1 + n_2 - 2} \tag{25.6}$$

The test statistic is then given by

$$t = \frac{\overline{X}_1 - \overline{X}_2}{s_p \sqrt{\dfrac{1}{n_1} + \dfrac{1}{n_2}}} \tag{25.7}$$

where t (Student's distribution) has $(n_1 + n_2 - 2)$ degrees of freedom. If the two unknown variances are assumed unequal, Equation 25.7 becomes

$$t = \frac{\overline{X}_1 - \overline{X}_2}{\sqrt{\dfrac{(s_1)^2}{n_1} + \dfrac{(s_2)^2}{n_2}}} \tag{25.8}$$

where t is now distributed with ν degrees of freedom. The term may be approximated from

$$\nu = \frac{\left[(s_1)^2 / n_1 + (s_2)^2 / n_2\right]^2}{\dfrac{\left[(s_1)^2 / n_1\right]^2}{n_1 + 1} + \dfrac{\left[(s_2)^2 / n_2\right]^2}{n_2 + 1}} - 2 \tag{25.9}$$

PROBLEMS AND SOLUTIONS

TMP 1: Large Sample Test of Hypothesis Concerning Mean Compressive Strength of Steel Beams

A random sample of 40 steel beams employed in the erection of an electrostatic precipitator has a mean compressive strength of 57,800 psi and a standard deviation of 650 psi. Test the hypothesis that the population of beams from which the sample was chosen has a mean, μ, equal to 58,000 psi against the alternative hypothesis that μ is less than 58,000 psi. Assume that the tolerated probability of Type I error is 0.05.

Solution

Identify the null hypothesis, H_0, and the alternative hypothesis, H_1:

$$H_0: \mu = 58{,}000 \text{ psi}$$

$$H_{1:} \mu < 58{,}000 \text{ psi}$$

Because the sample size, n, is 40, the large sample test is applicable and Equation 25.1 applies. The test statistic is therefore

$$Z = \frac{\overline{X} - 58000}{\dfrac{\sigma}{\sqrt{40}}}$$

The critical region may now be determined. Because H_1 is $\mu < \mu_0$, and α, the tolerated probability of a Type I error, is 0.05, the critical region is characterized by $Z < -Z_{0.05}$ (a one-sided test), where $Z_{0.05} = 1.645$. The critical region, therefore, is specified by

$$\frac{\overline{X} - 58000}{\dfrac{\sigma}{\sqrt{40}}} < -1.645$$

Because σ is unknown, it may be replaced by the sample standard deviation, $s = 650$ psi. Therefore, the observed calculated value of the test statistic is

$$\frac{57800 - 58000}{\dfrac{650}{\sqrt{40}}} = -1.95$$

Because -1.95 is less than -1.645, reject H_0: $\mu = 58,000$ psi and conclude that the population mean compressive strength is less than 58,000 psi.

TMP 2: Calculation of P-Values

Refer to Problem TMP 1. What is the P-value?

Solution

The P-values for various alternative hypotheses are shown in Table 25.3

The P-value is obtained by finding the probability of obtaining a test statistic value more extreme than the value actually observed, z_0, if H_0 is true. The value of the test statistic actually observed is $Z_0 = -1.95$. Because the alternative hypothesis, H_1, is $\mu < 58,000$, the P-value is $P(Z < -1.95)$ when $\mu = 58,000$. From the table of the standard normal distribution $P(Z < -1.95) = 0.0256$. The P-value of the test is 0.0256, which may be rounded off to 0.03 or 3%. The P-value provides an indication of the strength or confidence of the rejection of H_0. The farther the P-value is below, the greater the strength of rejection of H_0.

TABLE 25.3
P-Values

H_1	P-Value
$\mu < \mu_0$ (one tail)	$P(Z < z_0)$
$\mu > \mu_0$ (one tail)	$P(Z > z_0)$
$\mu \neq \mu_0$ (two tails)	$2P(Z > |z_0|)$

TMP 3: Estimating the True Mean

The weight fraction of carbon in a carbon nanotube impregnated with a specialty metal is estimated to be 0.875. Tests on 17 nanotube samples produced a mean weight fraction of 0.871 with a standard deviation of 0.0115. Test the hypothesis that the samples were drawn from a population with a mean weight fraction of 0.875. Set the level of significance at 5% (0.05).

Solution

The null and alternative hypotheses for this application are

$$H_0: \mu = 0.875$$

$$H_{1:} \mu \neq 0.875$$

First calculate t from s and $\overline{X}$ where

$$\overline{X} = 0.875$$

$$s = 0.0115$$

$$t = \frac{|\overline{X} - \mu|}{s / \sqrt{n}}$$

$$= \frac{|0.871 - 0.875|}{0.0115 / \sqrt{17}}$$

$$= 1.43$$

The corresponding tabulated value of t at 0.05 level with 16 degrees of freedom is for this two-tailed test

$$t_{0.025,16} = 2.12$$

Because the calculated value is less than the tabulated value, the hypothesis is accepted (not rejected).

Once again the results are as follows: with the scatter as reported in the standard deviation, one cannot really tell the difference between 0.871 and 0.875 from the tests.

Regarding the level of significance, the 0.05 level states that there is 1 chance in 20 that the true mean lies outside the specified range. At the 0.01 level, there is only 1 chance in 100. At the 0.1 level, there is 1 chance in 10.

With regard to the confidence range, written as

$$\mu = \overline{X} \pm ts$$

where the term s can be viewed as an estimation of the "precision" of the measurement of $\overline{X}$ and that 95% of similar measurements of X would fall within the range of $\overline{X} \pm ts$, when t is taken at the 0.05 significance level.

TMP 4: Small Sample Test of Hypothesis Concerning Mean Tar Content of Cigarettes

A cigarette manufacturer claims that the average tar content of its brand of cigarettes is 14.0 mg per cigarette. The tar content of a random sample of five cigarettes produced by this manufacturer is measured with the following results:

$$14.7, 14.1, 14.3, 14.5, 14.6 \text{ mg}$$

Do the results provide evidence that the average tar content of the manufacturer's cigarettes exceeds the value claimed? Assume cigarette tar content is normally distributed, and that the tolerated probability of a Type I error is 0.01.

Solution

The null hypothesis, H_0, and the alternative hypothesis, H_1, are as follows:

H_0: $\mu = 14.0$ where μ is the mean tar content in mg of the cigarettes produced by the manufacturer.

H_1: $\mu > 14.0$ states that the manufacturer's claim is false because the mean tar content exceeds that claimed by the manufacturer.

Note that the sample size is less than 30 (the sample size $n = 5$), no information is given about σ, and the problem states that the population sampled is normal. Therefore, the test statistic is now given by Equation 25.2:

$$t = \frac{\overline{X} - \mu_0}{\dfrac{s}{\sqrt{n}}}$$

The reader is left the exercise of showing that $\overline{X}$, the mean of the sample observations, is 14.44; s, the sample standard deviation, is 0.241. Therefore, with $n = 5$,

$$t = \frac{14.44 - 14.0}{\dfrac{0.241}{\sqrt{5}}}$$

$$= 4.08$$

Because the alternative hypothesis, H_1, is $\mu > 14.0$ (a one-sided test) and α is 0.01, the critical region is $t > t_{0.01}$. From the table of Student's distribution, $t_{0.01}$ for 4 degrees of freedom is 3.747. Therefore, the results of the test are as follows:

H_0: $\mu = 14.0$ is rejected because the observed value of the test statistic is 4.08, which is greater than 3.747. The sample evidence indicates that the manufacturer's claim is false and that the average tar content of cigarettes produced exceeds 14.0 mg.

TMP 5: Two-Tailed Thermometer Test

The mean lifetime of a sample of 100 thermometers produced by a manufacturer is determined to be 1620 weeks with a standard deviation of 100 weeks. If μ is the sample mean lifetime, test the hypothesis μ = 1600 weeks vs. the alternative hypothesis $\mu \neq$ 1600 weeks. Employ a level of significance $\alpha = 0.01$.

Solution

Set the null and alternative hypotheses:

$$H_0: \mu = 1600 \text{ weeks}$$

$$H_1: \mu \neq 1600 \text{ weeks}$$

Also note that a two-tailed test should be employed because the alternative hypothesis $\mu \neq$ 1600 weeks includes values both below and above 1600.

The variable under consideration is the thermometer sample mean $\overline{X}$. The distribution of $\overline{X}$ has a mean $\mu_{\overline{X}} = \mu$ and standard deviation $\sigma_{\overline{X}} = \sigma/\sqrt{n}$, where μ and σ are the mean and standard deviation of all thermometers produced by the company. Since σ is not known (or given), assume the standard deviation of the sample (of 100 thermometers) as a best-guess estimate of σ. For H_0,

$$\mu = 1600$$

$$\sigma_{\overline{X}} = \sigma/\sqrt{n} = 100/\sqrt{100}$$

$$= 10$$

Thus,

$$Z = \frac{1620 - 1600}{10} = 2.0$$

For a level of significance of 0.01 (two sided), the acceptable range is –2.58 to 2.58. Because it is inside the acceptable range, H_0 is accepted.

If the level of significance were 0.05, the acceptable range would become –1.96 to 1.96. Here, the null hypothesis (H_0 = 1600) is rejected because $Z > 1.96$, and therefore outside the acceptable range.

TMP 6: TYPE II ERROR CALCULATION

Refer to Problem TMP 1. What is the probability of a Type II error when the true value of μ is 57,600 psi?

Solution

To obtain the probability of a Type II error, one must determine the probability that H_0 is not rejected when H_0 is false. To obtain the probability of a Type II error, proceed as follows. The term β is the probability that H_0 is not rejected when H_0 is false. Therefore,

$$\beta = P\left(\frac{\overline{X} - 58,000}{\frac{650}{\sqrt{40}}} > -1.645 \right) \text{when } \mu < 58,000$$

TABLE 25.4
Thermometer Readings

Reading No.	Temperature (T) in °C		
	Thermometer 1 (T_1)	Thermometer 2 (T_2)	$\Delta T;\ (T_1 - T_2)$
1	63	63	0
2	51	50	1
3	67	62	5
4	56	60	-4
5	74	70	4
6	76	70	6
7	36	33	3

Solving the inequality for $\overline{X}$ produces the following expression for β:

$$\beta = P\left(\overline{X} > 58000 - 1.645 \frac{650}{\sqrt{40}}\right)$$

$$= P(\overline{X} > 57831)$$

Finding the value of the probability of a Type II error when $\mu = 57{,}600$ involves converting $P(\overline{X} > 57831)$ to a statement about a standard normal variable Z.

$$P(\overline{X} > 57831) = P\left(\frac{\overline{X} - 57600}{\frac{650}{\sqrt{40}}} > \frac{57831 - 57600}{\frac{650}{\sqrt{40}}}\right)$$

$$= P(Z > 2.25) = 0.0122$$

Therefore, the probability of a Type II error when $\mu = 57{,}600$ is approximately 0.01.

TMP 7: DIFFERENCE BETWEEN THE MEAN AND ZERO

Temperature readings from two different thermometers located in an aerobic digestor were recorded. Determine whether there is a significant difference between the thermometers at the 0.05 level. The data are provided in Table 25.4.

Solution

Key calculations are provided as follows:

$$\sum \Delta T = 15$$

$$\overline{\Delta T} = 2.143$$

$$s^2(\Delta T) = 1.687$$

$$s(\Delta T) = 1.300$$

The t test is employed for this sample of $n = 7$:

$$t = \frac{|2.143 - 0|}{1.300}$$

$$= 1.648$$

The observed or critical value t (two-tailed) is

$$t_{0.05,6} = 2.447$$

Because the calculated t is less than the critical value, the hypothesis that there is no difference between the thermometer readings is accepted.

TMP 8: ALTERNATE SIGNIFICANT LEVELS

Refer to Problem TMP 7. Repeat the calculation for the following two levels:

1. 0.10
2. 0.20

Solution

Obviously, the calculated value remains the same. The critical value does change for each of the two cases.

1. At 0.10,

$$t_{0.10,6} = 1.943$$

The hypothesis is again accepted.

2. At 0.20,

$$t_{0.20,6} = 1.440$$

The hypothesis is rejected because the calculated value is greater.

TMP 9: OPERATING TEMPERATURES OF TWO CRYSTALLIZERS

The daily operating temperature over a 50-d period for crystallizer A has a mean temperature of 17.0°C with a standard deviation of 2.5°C, whereas companion unit B recorded a mean of 17.8°C with a standard deviation of 3.0°C over a 70-d period. Can one conclude that unit B operates at a higher temperature than A at a 5% level of significance?

Solution

Because the standard deviations, σ_A and σ_B, of the two populations are again unknown, no serious error is committed by assuming that the standard deviation of each population is identical with the standard deviation of the sample taken from that population.

The following hypothesis is proposed:

$$H_0: \mu_A = \mu_B$$

The data given are

$$\overline{X}_A = 17, \quad \overline{X}_B = 17.8$$

$$\sigma_A = 2.5, \quad \sigma_B = 3.0$$

$$n_A = 50, \quad n_B = 70$$

Therefore,

$$\sigma_{\overline{A}} = \frac{\sigma_A}{\sqrt{n_A}} = \frac{2.5}{\sqrt{50}}$$

$$\sigma_{\overline{B}} = \frac{\sigma_B}{\sqrt{n_B}} = \frac{3.0}{\sqrt{70}}$$

From Equation 25.5,

$$\sigma_{\overline{A}-\overline{B}} = \sqrt{\sigma_{\overline{A}}^2 + \sigma_{\overline{B}}^2}$$

$$= \sqrt{\frac{(2.5)^2}{50} + \frac{(3.0)^2}{70}}$$

$$= 0.50$$

The value of Z may now be calculated:

$$Z = \frac{(\overline{X}_A - \overline{X}_B) - (\mu_A - \mu_B)}{\sigma_{\overline{A}-\overline{B}}}$$

$$= \frac{(17.0 - 17.8)}{0.50} - 0$$

$$= \frac{-0.80}{0.50} = -1.60$$

The solution in this problem is based on the calculated probability. The calculated probability for $Z = 1.60$ is approximately 10.96%. Because this is greater than the given value of 5%, the hypothesis is considered true; i.e., H_0 cannot be rejected. One concludes that both crystallizers operate at the same temperature.

TMP 10: COMPARISON OF TWO MEANS

An equipment manufacturer is marketing a new catalyst that they claim provides a greater yield in the conversion of waste. Seven runs are made with the catalyst, and their mean yield of 90.1% is to be compared with the yield of 87.7% for 7 runs of the present catalyst. The pooled estimate of the standard deviation from the 14 trials is 4.51 units. Determine whether the new catalyst is different from the present catalyst. Perform the test at the 0.1 level.

Solution

The null hypothesis in this case is given by

$$H_0: \mu_1 = \mu_2$$

The calculated value of t is first obtained. From Equation 25.7,

$$t = \frac{\overline{X}_1 - \overline{X}_2}{s_p \sqrt{\dfrac{1}{n_1} + \dfrac{1}{n_2}}}$$

Based on the problem statement,

$$\overline{X}_1 = 90.1$$

$$\overline{X}_2 = 87.7$$

$$s_p = 4.52$$

$$n_1 = 7$$

$$n_2 = 7$$

Substituting,

$$t = \frac{90.1 - 87.7}{4.52 \sqrt{\dfrac{1}{7} + \dfrac{1}{7}}}$$

$$= 1.14$$

The critical or tabulated value of t is (with 12 degrees of freedom)

$$t_{0.10,12} = 1.356$$

Because the critical value of t is greater than the calculated value, the null hypothesis is accepted. In effect, the population mean for X_1 is not significantly different from X_2, with 0.1 chance of being wrong.

TMP 11: CHOLESTEROL COUNT PROPORTION

Cholesterol counts from 50 males yielded results that 39 were above an acceptable number. The same test on 60 women indicated that 36 were in the unacceptable range. Do these data suggest that the count of men is greater than that of women? Perform the analysis employing a 5% level of significance.

Solution

The proportion of men (M) who exceeded the acceptable level is

$$p_M = \frac{39}{50} = 0.78$$

For women (W),

$$p_W = \frac{36}{60} = 0.60$$

The hypotheses for this case are

$$H_0: p_M = p_W$$

$$H_1: p_M > p_W$$

For H_0,

$$p_M - p_W = 0$$

and for a binomial distribution (see Chapter 23),

$$\sigma_{p_M - p_W} = \sqrt{pQ[(1/50) + (1/60)]}$$

where

$$p = (39 + 36)/110 = 0.682$$

$$Q = 1 - p = 0.318$$

Thus,

$$\sigma_{p_M - p_W} = \sqrt{(0.682)(0.318)[(1/50) + (1/60)]}$$

$$= 0.0892$$

The calculated value of Z is

$$Z = \frac{0.78 - 0.60}{0.0892}$$

$$= 2.018$$

This value is greater than the critical value of 1.645 α at = 0.05 (one-tailed test). Because

$$2.018 > 1.645$$

one rejects the hypothesis; i.e., the test violates the hypothesis H_0.

TMP 12: ONE-TAILED HYPOTHESIS FOR PUMPS

Past experience at a manufacturing plant indicates that 10% ($d = 0.1$) of the pumps sold are defective in some manner or form. A recent test of 400 pumps resulted in 50 pumps being defective. Does the recent test result indicate that the hypothesis based on past experience is no longer valid (defectives are more than 10%) at the 10% level of significance?

Solution

The null hypothesis is

$$H_0: d = 0.1$$

The alternative hypothesis is

$$H_1: d > 0.1$$

Assuming a normal distribution, the random variable, Z, is again given by

$$Z = \frac{d_s - d}{\sigma}$$

with zero mean and a standard deviation of unity. Thus, for this one-tailed application,

$$P(Z > Z_{0.1}) = 0.10$$

with

$$Z_{0.1} = 1.282$$

Thus, if H_0 is true, Z must take on a value less than 1.282.
Assume the binomial distribution applies; for a binomial distribution (see Chapter 23),

$$\sigma^2 = \frac{d(1-d)}{n}$$

$$= \frac{0.1(1-0.1)}{400}$$

$$= 0.000225$$

and

$$\sigma = 0.015$$

The experimental value of the sample proportion is

$$d_s = \frac{50}{400} = 0.125$$

The corresponding value of Z is

$$Z = \frac{0.125 - 0.1}{0.015}; \quad d = 0.1$$

$$= 1.667$$

This value is greater than the "allowed" upper limit of Z. Therefore, the hypothesis, H_0, is rejected at the 10% level of significance.

26 CHI. Chi-Square Distribution

INTRODUCTION

A random variable X is said to have a chi-square distribution with ν degrees of freedom if its probability distribution function (pdf) is specified by

$$f(\chi^2) = \frac{1}{2^{\frac{\nu}{2}}\Gamma\left(\frac{\nu}{2}\right)}(\chi^2)^{\frac{\nu}{2}-1}e^{\frac{-\chi^2}{2}}; \quad \chi^2 > 0; \quad \Gamma = \text{Gamma function} \tag{26.1a}$$

or

$$f(\chi) = \frac{1}{2^{\frac{\nu}{2}}\Gamma\left(\frac{\nu}{2}\right)}\chi^{\frac{\nu}{2}-1}e^{\frac{-\chi}{2}}; \quad \chi > 0; \quad \Gamma = \text{Gamma function} \tag{26.1b}$$

where χ is the Greek letter chi (chi is pronounced as in kite), and χ^2 is read as chi-square. The above equation may also be written as

$$f(\chi^2) = A(\chi^2)^{\frac{\nu}{2}-1}e^{\frac{-\chi^2}{2}} \tag{26.2}$$

Here, $\nu = n - 1$ is once again the number of degrees of freedom, and A is a constant depending on a value of ν such that the total area under the curve is unity. The chi-square distributions corresponding to various values of ν are shown in Figure 26.1. The maximum value of $f(\chi^2)$ occurs at $\chi^2 = \nu - 2$ for $\nu \geq 2$.

For random samples from a normal population with variance σ^2, the statistic

$$\chi^2 = \frac{(n-1)s^2}{\sigma^2} \tag{26.3}$$

where s^2 is the sample variance, and n, the sample size, has a chi-square distribution with $n - 1$ degrees of freedom. The variance may be calculated using the procedures outlined in Chapter 9.

A test of the hypothesis H_0: $\sigma^2 = \sigma_0^2$ utilizes the test statistic

$$\frac{(n-1)s^2}{\sigma_0^2} \tag{26.4}$$

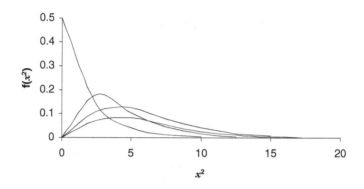

FIGURE 26.1 Chi-square distributions for various values of v.

The critical region is determined by the alternative hypothesis, H_1, and α, the tolerated probability of a Type I error.

As with the normal and t distributions, one can define 95%, 99%, or other confidence limits and intervals for χ^2 by use of the table of the χ^2 distribution (see Table A.3, Appendix A). In this manner, one can estimate within specified limits of confidence the population standard deviation σ in terms of a sample standard deviation, s. For example, $\chi^2_{.025}$ and $\chi^2_{.975}$ are the values of χ^2 (called critical values) for which 2.5% of the area lies in each "tail" of the distribution, then the 95% confidence interval is

$$\chi^2_{.025} < \frac{(n-1)s^2}{\sigma^2} < \chi^2_{0.975}$$

from which one can see that σ is estimated to lie in the interval

$$\frac{s\sqrt{n-1}}{\chi_{0.975}} < \sigma < \frac{s\sqrt{n-1}}{\chi_{0.025}}$$

with 95% confidence. Similarly, other confidence intervals can be found. The values $\chi_{0.025}$ and $\chi_{0.975}$ represent, respectively, the 2.5 and 97.5 percentile values. The preceding two equations may more appropriately be written as

$$\chi^2_{0.025} < \frac{(n-1)s^2}{\sigma^2} \quad \text{or} \quad \chi^2_{0.975} > \frac{(n-1)s^2}{\sigma^2}$$

and

$$\sigma < \frac{s\sqrt{n-1}}{\chi_{0.025}} \quad \text{or} \quad \sigma > \frac{s\sqrt{n-1}}{\chi_{0.975}}$$

Table A.3 (Appendix A) gives percentile values corresponding to the number of degrees of freedom. For large values of v ($v \geq 30$), one can use the fact that $(\sqrt{2\chi^2} - \sqrt{2v-1})$ is very nearly normally distributed with mean 0 and standard deviation 1 so that normal distribution tables can be used (if $v \geq 30$). Then if χ^2_p and Z_p are the Pth percentiles of the chi-square and normal distributions, respectively, one has

$$\chi_p^2 = \frac{1}{2}(Z_p + \sqrt{2v-1})^2 \tag{26.5}$$

PROBLEMS AND SOLUTIONS

CHI 1: CHI-SQUARE APPLICATION

Provide applications of the chi-square distribution.

Solution

The chi-square distribution is of great practical and theoretical interest. It is used in the following areas:

1. To calculate a confidence interval from sample data for the variance of a normal distribution
2. To calculate a confidence interval from sample data for the parameter of an exponential distribution
3. To conduct various "goodness of fit" tests to determine whether a given sample could reasonably have come from a normal distribution
4. To determine whether significant differences exist among the frequencies in a contingency table

Occasionally, the chi-square distribution has provided a reasonable approximation of the distribution of certain physical variables. For example, it has represented compressor frequency of large jet engines by a chi-square distribution with 5 degrees of freedom.

CHI 2: CHI-SQUARE TABLE APPLICATION

Answer the following two questions:

1. For 2 degrees of freedom ($v = 2$), what is the probability of obtaining a χ^2 value equal to or greater than 4.605?
2. For 20 degrees of freedom, what value of χ^2 will provide a 5% (0.05) probability that χ^2 is equal to or larger than this value?

Solution

From Table A.3 (Appendix A),

1. Probability = 0.10 = 10%
2. $\chi^2 = 31.41$

CHI 3: CHI-SQUARE VALUES

Consider a chi-square distribution with 9 degrees of freedom. Find the critical values for $\chi^2_{0.05}$ and $\chi^2_{0.95}$. Also obtain the critical values for $\chi^2_{0.05}$ and $\chi^2_{0.95}$ if the degrees of freedom is 14.

Solution

Refer to Table A.3 (Appendix A). For $v = 9$,

$$\chi^2_{0.05} = 16.919$$

and

$$\chi^2_{0.95} = 3.325$$

For $\nu = 14$,

$$\chi^2_{0.05} = 23.685$$

and

$$\chi^2_{0.95} = 6.571$$

CHI 4: CHI-SQUARE CONFIDENCE INTERVAL

Obtain the 90% confidence interval for χ^2 if $\alpha = 0.05$ with $\nu = 19$.

Solution

For this case, refer to Equation 26.3.

$$\chi^2_{0.95} < \chi^2 < \chi^2_{0.05}$$

$$\chi^2_{0.95} < \frac{(n-1)s^2}{\sigma^2} < \chi^2_{0.05}$$

From Table A.3 (Appendix A), one obtains

$$10.117 < \frac{(n-1)s^2}{\sigma^2} < 30.144$$

If the sample standard deviation, s, was specified, the confidence interval for the population standard deviation would be given by

$$\frac{s\sqrt{n-1}}{\sqrt{30.114}} < \sigma < \frac{s\sqrt{n-1}}{\sqrt{10.117}}$$

CHI 5: TESTING HYPOTHESES CONCERNING VARIABILITY OF GAS MILEAGE

A car manufacturer reports to the United States Environmental Protection Agency (USEPA) that the gas mileage in miles per gallon (mpg) for a new model has a mean equal to 38 and a standard deviation equal to 3. A consumer service tests a random sample of 15 cars of the new model and obtains the following gas mileages:

37 39 42 45 34 32 36 36 38 43 40 43 37 30 38

Assuming gas mileage is normally distributed, test the hypothesis H_0: $\sigma = 3$ against the alternative hypothesis H_1: $\sigma > 3$. Use 0.05 as the tolerated probability of a Type I error.

Solution

Note that the gas mileage is assumed to be normally distributed. The one-tail critical region statistic is

$$\frac{(n-1)s^2}{\sigma_0^2} > \chi^2_{0.05}$$

For H_0: $\sigma = 3$ the value of σ_0 is 3. From the table of the chi-square distribution, $\chi^2_{0.05} = 23.7$ with $\nu = 14$. Therefore, the critical region is specified by

$$\frac{(n-1)s^2}{9} > 23.7$$

The value of the test statistic $\dfrac{(n-1)s^2}{\sigma_0^2}$ may now be calculated. For the given data, $n = 15$ and, as can be calculated using the procedures set forth in Chapter 9, $s = 4.19$. Therefore, the value of the test statistic is

$$\frac{(14)(4.19)^2}{9} = 27.3$$

The hypothesis test may now be completed. Reject H_0: $\sigma = 3$ because the observed value of the test statistic, 27.3, exceeds 23.7. Thus, the data do not support the manufacturer's claim that the standard deviation of gas mileage for the new model is 3 mpg. The standard deviation exceeds 3 mpg by a significant amount.

CHI 6: VARIABLE PROBABILITY OF TYPE I ERROR

Refer to Problem CHI 5. Repeat the calculation if the tolerated probability of a Type I error is

1. 0.025
2. 0.01

Solution

The value of the test statistic

$$\frac{(n-1)s^2}{\sigma_0^2}$$

remains the same at 27.3. The critical region, however, does change.

1. For a tolerated probability of a Type I error of 0.025, the critical region is specified as

$$\frac{(n-1)s^2}{9} > \chi^2_{0.025}$$

$$> 26.1$$

Because the observed value of the test statistic 27.3 exceeds 26.1, the hypothesis is again rejected.

2. For a tolerated probability of a Type I error of 0.01, the critical region is specified as

$$\frac{(n-1)s^2}{9} > \chi^2_{0.01}$$

$$> 29.1$$

Because the test statistic is below 27.3, this hypothesis is accepted.

CHI 7: PROBABILITY CALCULATION

Calculate the probability that 13 atmospheric pressure readings (mbar) from a population with variance of 0.8 will have a sample variance greater than 1.6.

Solution

Assume the population pressure readings are normally distributed. As before, the random variable

$$\frac{(n-1)s^2}{\sigma^2}$$

has a chi-square distributed with $n-1$ or degrees of freedom. The describing equation is written as

$$P(s^2 > 1.6) = \left(\frac{(n-1)s^2}{\sigma^2} > \frac{12(1.6)}{0.8} \right)$$

$$= P(\chi^2 > 24)$$

From Table A.3 (Appendix A), with $\nu = 12$,

$$P(s^2 > 1.6) = 0.021 \text{ (linear interpolation)}$$

CHI 8: GOODNESS-OF-FIT TESTING PROCEDURE

Outline a procedure that employs the chi-square distribution in testing goodness of fit.

Solution

One of the principal applications of the chi-square distribution is testing goodness of fit. A test of goodness of fit is a test of the assumption that the pdf of a random variable has some specified form. The chi-square test of goodness of fit utilizes as a test statistic

$$\sum_{i=1}^{k} \frac{(f_i - e_i)^2}{e_i} \qquad (26.6)$$

where k is the number of categories in which the observations on the random variable have been classified, f_i is the observed frequency of the ith category, and e_i is the corresponding theoretical frequency; $i = 1, \ldots, k$ under the assumption being tested. If the assumption being tested is true, then the chi-square test statistic is approximately distributed as a random variable having a chi-square distribution with degrees of freedom $v = k - m - 1$, where m is the number of parameters estimated from the data in order to obtain the theoretical frequencies for each of the k categories in which the data have been classified. Large values of the chi-square test statistic lead to rejection of the assumption being tested.

The P-value of the test is the probability that a random variable having a chi-square distribution with v degrees of freedom exceeds the calculated value of the chi-square test statistic. When the P-value is small, usually less than 0.05, then the assumption being tested is rejected. This procedure is detailed in Problem CHI 9.

CHI 9: NUCLEAR FACILITY PUMP FAILURE

A nuclear facility employs a large number of pumps: 30% are of design I and 70% of design II. Ten years later, there have been 128 pump failures of design I and 312 of design II. Comment on whether there is evidence of a difference between the two designs.

Solution

If there were no difference between the designs, the expected number of failures would be proportionally the same. For this application,

$$\text{Failures, I} = (0.30)(440)$$

$$= 132$$

$$\text{Failures, II} = (0.70)(440)$$

$$= 308$$

Employ Equation 26.6 to calculate χ^2:

$$\chi^2 = \frac{(128-132)^2}{132} + \frac{(312-308)^2}{308}$$

$$= 0.121 + 0.052$$

$$= 0.173$$

Because there are only two designs, there is only 1 degree of freedom. From Table A.3 (Appendix A), one notes that with 1 degree of freedom, there is only a 0.007 (0.70%) probability that χ^2 would be larger than 0.173. Therefore, the probability that there is no difference between the two pump designs is less than a 0.007.

The reader should note that the expected frequency does not occur every time. If one tosses a fair die 12 times, one would not expect to see 2 ones, 2 twos, ..., 2 sixes. Alternately, if the mean pump failure in a plant is 1 every 30 months, one would not expect 10 pumps to fail every 300 months, although 1 in 30 is the "expected" frequency.

CHI 10: TESTING ASSUMPTION CONCERNING THE PDF OF INTERREQUEST TIMES FOR COMPUTER SERVICE

Table 26.1 shows the observed frequency distribution and theoretical frequency distribution of 50 interrequest times for computer service. The theoretical frequencies were obtained under the

TABLE 26.1
Frequency of Interrequest Time

Interrequest Time (μsec)	Observed Frequency	Theoretical Frequency
0–2499	5	5.6
2500–4999	11	10.8
5000–9999	18	14.8
10000–19999	10	11.8
20000–39999	5	5.5
40000–79999	0	1.4
80000–159999	1	0.3

assumption that interrequest time for computer service has a log-normal distribution. Refer to Problem LOG 3 in Chapter 19 for additional details.

On the basis of the given data, test the assumption that interrequest time for computer service has a log-normal distribution. Also determine the number of parameters estimated from the data.

Solution

The number of classes in the frequency distribution, k, is 7. There are two parameters that can be estimated from the data, μ and σ. Therefore, $m = 2$.

The degree of freedom, ν, is therefore

$$\nu = k - m - 1$$

$$= 7 - 2 - 1 = 4$$

The chi-square test statistic is now evaluated employing Equation 26.8.

$$\sum_{i=1}^{k} \frac{(f_i - e_i)^2}{e_i}$$

$$= \frac{(5-5.6)^2}{5.6} + \frac{(11-10.8)^2}{10.8} + \frac{(10-11.8)^2}{11.8} + \frac{(5-5.5)^2}{5.5} + \frac{(0-1.4)^2}{1.4} + \frac{(1-0.3)^2}{0.3}$$

$$= 4.83$$

As indicated in Problem CHI 9, the probability that a chi-square variable with ν degrees of freedom exceeds the observed value of the test statistic is given by

$$P(\chi^2 > 4.83) > 0.30$$

If the probability computed is less than 0.05, one would reject the assumption being tested; otherwise, one would fail to reject it. In this case, one would fail to reject the assumption that the interrequest time for computer service has a log-normal distribution. Note that rejecting the assumption being tested indicates that the data do not provide sufficient evidence in favor of the assumed distribution of the random variable.

CHI 11: CONTINGENCY TABLE PROCEDURE

Provide a procedure that may be employed to test the independence of the criteria of classification.

Solution

A contingency table is a classification of observations according to two or more criteria of classification. An $r \times c$ contingency table features r rows and c columns. Contingency tables are used to test the independence of the criteria of classification. The observed frequency of each cell, i.e., the intersection of a row and column, is compared with the corresponding theoretical frequency. Under the assumption that the row and column criteria are independent, multiplying the row total by the column total and then dividing the result by the grand total leads to the theoretical frequency of each cell. The chi-square test statistic is then computed. Under the assumption of independence, the chi-square test statistic is approximately distributed as a random variable having a chi-square distribution with $(r-1)(c-1)$ degrees of freedom. Large values of the chi-square test statistic lead to rejection of the assumption of independence. This procedure is demonstrated in Problem CHI 12.

CHI 12: INDEPENDENCE OF WORK SHIFT AND QUALITY OF OUTPUT

A sample of 300 items from one day's production of a factory is classified as to work shift and quality as follows:

Quality	Day	Evening	Midnight	Total
Poor	12	12	10	34
Satisfactory	90	68	72	230
Excellent	21	10	5	36
Total	123	90	87	300

Test the assumption that quality of items produced is independent of the work shift in which they are produced. Use a 10% level of significance.

Solution

The product of the row total and the column total divided by the grand total is computed for each cell:

$$e_{11} = \frac{(34)(123)}{300} = 13.9$$

$$e_{12} = \frac{(34)(90)}{300} = 10.2$$

$$e_{13} = \frac{(34)(87)}{300} = 9.9$$

$$e_{21} = \frac{(230)(123)}{300} = 94.3$$

$$e_{22} = \frac{(230)(90)}{300} = 69.0$$

$$e_{23} = \frac{(230)(87)}{300} = 66.7$$

$$e_{31} = \frac{(36)(123)}{300} = 14.8$$

$$e_{32} = \frac{(36)(90)}{300} = 10.8$$

$$e_{33} = \frac{(36)(87)}{300} = 10.4$$

Note that e_{ij} represents the theoretical frequency for the cell at the intersection of the ith row and jth column.

The test statistic is given by

$$\sum_{i=1}^{r} \sum_{j=1}^{c} \frac{(f_{ij} - e_{ij})^2}{e_{ij}}$$

where f_{ij} is the observed frequency of the ith row and the jth column, i.e., the cell at the intersection of the ith row and the jth column, e_{ij} is the corresponding theoretical frequency, r is the number of rows, and c is the number of columns. Therefore, the chi-square test statistic is

$$\chi^2 = \frac{(12-13.9)^2}{13.9} + \frac{(12-10.2)^2}{10.2} + \frac{(10-9.9)^2}{9.9} + \frac{(90-94.3)^2}{94.3}$$

$$+ \frac{(68-69)^2}{69} + \frac{(72-66.7)^2}{66.7} + \frac{(21-14.8)^2}{14.8} + \frac{(10-10.8)^2}{10.8} + \frac{(5-10.4)^2}{10.4}$$

$$= 6.67$$

The probability (P-value) that a chi-square variable with $(r-1)(c-1)$ degrees of freedom exceeds the value of the chi-square test statistic is

$$P(\chi^2_{(r-1)(c-1)} > 6.67) > 0.10; \quad r = 3, c = 3, \quad (r-1)(c-1) = (2)(2) = 4$$

$$P(\chi^2_4 > 6.67) > 0.10$$

Because the P-value exceeds 0.10, one would fail to reject the assumption that the quality of an item is independent of the work shift during which it was produced. Note that rejection of independence implies that the row and column criteria of classification are interrelated.

27 FFF. The *F* Distribution

INTRODUCTION

As noted several times earlier, variance measures variability. Comparison of variability, therefore, involves comparison of variance. Suppose σ_1^2 and σ_2^2 represent the unknown variances of two independent normal populations. The null hypothesis, H_0: $\sigma_1^2 = \sigma_2^2$, asserts that the two populations have the same variance and are therefore characterized by the same variability. When the null hypothesis is true, the test statistic, s_1^2/s_2^2, has an *F* distribution with parameters $n_1 - 1$ and $n_2 - 1$ called degrees of freedom. Here, s_1^2 is the sample variance of a random sample of n_1 observations from the normal population having variance σ_1^2, while s_2^2 is the sample variance of a random sample of n_2 observations from an independent normal population having variance σ_2^2. When s_1^2/s_2^2 is used as a test statistic, the critical region depends on the alternative hypothesis, H_1, and α, the tolerated probability of a Type I error. Listed in Table 27.1 are the critical regions for three different alternative hypotheses.

The term F_α is the value which a random variable having an *F* distribution with $n_1 - 1$ and $n_2 - 1$ degrees of freedom exceeds with probability α. The terms $F_{\alpha/2}$ and $F_{1-\alpha/2}$ are defined similarly. Also note that

$$F_{\alpha;v_1,v_2} = \frac{1}{F_{1-\alpha;v_2,v_1}} \tag{27.1}$$

where v_1 and v_2 are again the degrees of freedom. Tabulated values of the *F* distribution can be found in Table A.4A and Table A.4B (Appendix A). In addition, the ratio s_1^2/s_2^2 must, of necessity, be greater than unity, i.e., $s_1^2/s_2^2 > 1.0$.

PROBLEMS AND SOLUTIONS

FFF 1: Distribution Background

Provide background information on the *F* distribution.

Solution

Student's t distribution was discussed previously in Chapter 25. With regard to means, it was noted that samples drawn from the same source or population, or measurements of other types, could not be expected to be identical. Obviously, the means of several samples vary over a range and this range can be approximated for any desired probability level. Similar comments apply for variances. As with means, if several groups of samples are obtained, the calculated variance of the population source obtained from each sample, once again will not be the same.

The *F* test and *F* distribution provide a method for determining whether the ratio of two variances is larger than might be normally expected by chance if the samples came from the same source or population. The procedure to follow is similar to that for comparing two means.

TABLE 27.1
Critical Regions: *F* Distribution

Alternative Hypothesis
(*H*₁)   **Critical Region**

$\sigma_1^2 \neq \sigma_2^2$	$s_1^2/s_2^2 < F_{1-\alpha/2}$ or $s_1^2/s_2^2 > F_{\alpha/2}$
$\sigma_1^2 > \sigma_2^2$	$s_1^2/s_2^2 > F_\alpha$
$\sigma_1^2 < \sigma_2^2$	$s_1^2/s_2^2 < F_\alpha$

The null hypothesis for comparing two variances is

$$H_0; \quad \sigma^2(X_1) = \sigma^2(X_2) \tag{27.2}$$

In the normal distribution and *t* test for the difference between two means, the difference between the two observed means was calculated and hypothesized to have come from populations with the same means. There is no statistic yet discovered that provides for the difference between two estimated variances. However, R.A. Fisher discovered a distribution of the *F* statistic given in Equation 27.2 and corresponding to the function $1/2 \log_e [s^2(X_2)/s^2(X_1)]$. Snedecor modified this to give the values corresponding to $s^2(X_2)/s^2(X_1)$ and named the ratio *F* in honor of Fisher. Therefore,

$$F = \frac{s^2(X_2)}{s^2(X_1)} \tag{27.3}$$

The terms $s^2(X_1)$ and $s^2(X_2)$ are the variances of the variables X_1 and X_2, calculated from n_1 measurements of X_1 and n_2 measurements of X_2. From Equation 27.3, one notes that *F* values have been tabulated in terms of an assumed level and two degrees of freedom:

$$\nu_1 = n_1 - 1$$

$$\nu_2 = n_2 - 1$$

This test provides information on whether one variance is larger than another, not whether two variances are significantly different. This effectively corresponds to the one-sided test employed in the comparison of means. To use the tables for a two-sided test, the indicated probability levels must be doubled.

FFF 2: *F* DISTRIBUTION APPLICATION

Discuss some *F* distribution applications.

Solution

Like the chi-square distribution, the *F* distribution is of importance in problems of statistical inference. It is used, for example, as follows:

1. To test whether, on the basis of sample data, there is evidence of a significant difference in the variance of two normally distributed populations.
2. To test whether, on the basis of sample data, there are significant differences among the averages of a number of normal populations. (This test is known as the analysis of variance, or ANOVA.)
3. To evaluate significance in least-squares regression analyses. In these applications, the parameters v_1 and v_2 are generally referred to as the degrees of freedom of the *F* distribution.

FFF 3: Obtaining *F* Distribution Values

Find the following *F* values:

1. $F_{0.01;5,3}$
2. $F_{0.95;3,12}$

Solution

This problem involves the use of Table A.4A and Table A.4B (Appendix A).

1. From Table A.4A:

$$F_{0.01;5,3} = 28.2$$

2. From Table A.4B:

$$F_{0.05;3,12} = 3.49$$

From Equation 27.1,

$$F_{0.95;3,12} = \frac{1}{F_{0.05;12,3}}$$

$$= \frac{1}{8.74}$$

$$= 0.114$$

FFF 4: Estimated Variance

A variance is known to be 0.417. A second variance is estimated to be 0.609 from 41 samples. Is the estimated variance significantly larger than the known variance of 0.417? Perform the test at both the 0.01 and 0.05 levels.

Solution

The key statistic is calculated from Equation 27.3. Note once again that the larger variance must be placed in the numerator.

$$F = \frac{s^2(X_2)}{s^2(X_1)}$$

$$= \frac{0.609}{0.417}$$

$$= 1.46$$

The tabulated values are obtained from Table A.4A and Table A.4B (Appendix A) by first noting that $v_1 = \infty$.

1. For the 0.01 level,

$$F_{0.01;40,\infty} = 1.59$$

Because the tabulated value is greater than the calculated statistic, the hypothesis that the two variances are equal is accepted; i.e., accept H_0.

2. For the 0.05 level,

$$F_{0.05;40,\infty} = 1.40$$

The hypothesis at this level is rejected.

From the preceding results, one can conclude that there is less than a 5% chance (or 0.05 probability) but more than a 1% chance that the variances came from the same population.

FFF 5: Weight Measurements

Twenty-five weight measurements from an analytical balance produce results for the sample variance of $s_1^2 = 0.45$ mg. Thirty-one weight measurements with another analytical balance produced a sample variance $s_2^2 = 0.20$ mg. Does the first balance have a σ_1 greater than σ_2? Test this hypothesis at the 1% level of significance.

Solution

For this case,

$$H_0: \sigma_1^2 = \sigma_2^2$$

$$H_1: \sigma_1^2 > \sigma_2^2$$

This is a one-sided (right tail) test. Noting that $\alpha = 0.01$, $v_1 = 24$, and $v_2 = 30$, Table A.4A (Appendix A) yields

$$F_{0.01;24,30} = 2.47$$

The critical region for the test is obtained from Table 27.1.

TABLE 27.2
PCB Data

Method	Sample Size	Standard Deviation (ppm)
1	8	260
2	12	201

$$\frac{s_1^{\,2}}{s_2^{\,2}} = \frac{0.45}{0.20}$$

$$= 2.25$$

Because 2.25 is not greater than 2.47, i.e., $2.25 < 2.47$, the hypothesis H_0 is accepted.

FFF 6: POLYCHLORINATED BIPHENYL (PCB) CONCENTRATION

The data in Table 27.2 are provided for two methods of measuring for the PCB concentration at a Superfund site.

Do the variances produced by the two methods differ significantly? Employ a 10% level of significance.

Solution

F calculated for the two methods is

$$F = \frac{s_1^2}{s_2^2} = \frac{(260)^2}{(201)^2}; \quad s_1 = 260, \; s_2 = 201$$

$$= 1.67$$

At the 10% level of significance with $v_1 = 7$ and $v_2 = 11$, one obtains (because this is a two-sided test)

$$F_{0.05;7,11} = 3.01 \text{ (by interpolation)}$$

Because $3.01 > 1.67$, the difference between the variances is not significant.

FFF 7: CRITICAL REGION

Sample variance of yield (in percentage) from two reactors is provided in Table 27.3.

If

$$H_0: \sigma_1^2 = \sigma_2^2$$

$$H_1: \sigma_1^2 > \sigma_2^2 \quad \text{or} \quad \sigma_1^2 < {}_2^2$$

obtain the critical region at a 10% level of significance.

TABLE 27.3
Reactor Data

Reactor	Sample Size	Sample Variance (s^2)
1	8	9.9
2	8	3.8

Solution

Note that F is distributed as

$$F_{7,7}$$

The critical region for this two-tailed test is therefore

$$F < F_{0.05;7,7} \quad \text{and} \quad F > F_{0.95;7,7}$$

From Table A.4B (Appendix A),

$$F < 3.79 \quad \text{and} \quad F > 0.264$$

For the given test,

$$F = \frac{9.9}{3.8} = 2.60$$

Because $0.264 < 2.60 < 3.79$, H_0 is accepted. The reader is left the exercise of showing that H_0 would be rejected at the 2% level of significance ($F = 6.99$).

FFF 8: PROBABILITY CALCULATION

The following data are provided on two sets, $n_1 = 10$, $n_2 = 15$, of soil samples from populations with equal variances, i.e., $\sigma_1 = \sigma_2$. Calculate

$$P(s_1^2/s_2^2 < 2.65)$$

Solution

For this one-sided right-handed test,

$$P(s_1^2/s_2^2 < 2.65) = P(F < 2.65)$$

with $v_1 = 9$ and $v_2 = 14$. From Table A.4B (Appendix A),

$$P(F < 2.65) = 1 - 0.05 = 0.95$$

TABLE 27.4
Battery Life Data

	Brand I	Brand II
Sample size	$m_1 = 8$	$n_2 = 10$
Sample mean	$\overline{X}_1 = 5$	$\overline{X}_2 = 5$
Sample variance	$s_1 = 2$	$s_2 = 1.2$

FFF 9: COMPARISON OF VARIABILITY OF BATTERY LIFE FOR TWO BRANDS OF BATTERIES

Two brands of automobile batteries are to be compared with respect to variability of battery life. For this purpose a random sample of eight batteries of Brand I and a random sample of 10 batteries of Brand II are compared with the results in Table 27.4.

Do the brands differ significantly with respect to variability? Assume the populations sampled are normal and independent, and that the tolerated probability of a Type I error is 0.02.

Solution

First identify the null hypothesis, H_0, and the alternative hypothesis, H_1:

$$H_0: \sigma_1^2 = \sigma_2^2$$

$$H_1: \sigma_1^2 = \sigma_2^2$$

where σ_1^2 is the population variance of Brand I and σ_2^2 is the population variance of Brand II.

Because the alternative hypothesis is $H_1: \sigma_1^2 \neq \sigma_2^2$ and $\alpha = 0.02$ (a two-sided test), the critical region is defined by

$$\frac{S_1^2}{S_2^2} < F_{0.99} \quad \text{or} \quad \frac{S_1^2}{S_2^2} > F_{0.01}$$

with

$$n_1 - 1 = 7$$

$$n_2 - 1 = 9$$

For a random variable having an F distribution with 7 and 9 degrees of freedom, $F_{0.01} = 5.61$, i.e., the value exceeded with probability of 0.01 by a random variable having an F distribution with 7 and 9 degrees of freedom. As described earlier, $F_{0.99}$ is obtained by making use of the fact that a random variable having an F distribution with 7 and 9 degrees of freedom is the reciprocal of a random variable having an F distribution with 9 and 7 degrees of freedom. Let $F_{7,9}$ and $F_{9,7}$ represent the original and reciprocal random variables, respectively. One then obtains from Equation 27.1,

$$F_{0.99;7,9} = \frac{1}{F_{0.01;9,7}}$$

$$= \frac{1}{6.72} = 0.15$$

The critical region, therefore, consists of those values of the test statistic, s_1^2 / s_2^2, exceeding 5.61 or less than 0.15. The test statistic may now be calculated:

$$\frac{s_1^2}{s_2^2} = \frac{2}{1.2}$$

$$= 1.67$$

Because 1.67 lies between 0.15 and 5.61, H_0 is not rejected. Therefore, Brand I and Brand II do not differ significantly with respect to variability of battery life.

FFF 10: VARIABLE TYPE I ERROR

Refer to Problem FFF 9. Do the brands differ significantly with respect to variability at a tolerated probability of a Type I error equal to 0.10?

Solution

With the change in the Type I error, the critical region changes although the test statistic remains the same. For this case,

$$F_{0.05;7,9} = 3.29$$

and

$$F_{0.95;7,9} = \frac{1}{F_{0.05;9,7}}$$

$$= \frac{1}{3.68} = 0.272$$

Surprisingly, the H_0 is again not rejected. A much larger tolerated probability would need to be specified for H_0 to be rejected.

28 REG. Regression Analysis: Method of Least Squares

INTRODUCTION

It is no secret that many statistical calculations are now performed with spreadsheets or packaged programs. This statement is particularly true with regression analysis. The result of this approach has been to often reduce or eliminate one's fundamental understanding of this subject. This chapter attempts to correct this shortcoming.

Engineers and scientists often encounter applications that require the need to develop a mathematical relationship between data for two or more variables. For example, if Y (a dependent variable) is a function of or depends on X (an independent variable), i.e.,

$$Y = f(X)$$

one may require to express this (X, Y) data in equation form. This process is referred to as *regression analysis*, and the regression method most often employed is the method of *least squares*.

An important step in this procedure — which is often omitted — is to prepare a plot of Y vs. X. The result, referred to as a scatter diagram, could take on any form. Three such plots are provided in Figure 28.1 (A to C).

The first plot (A) suggests a linear relationship between X and Y, i.e.,

$$Y = a_0 + a_1 X$$

The second graph (B) appears to be best represented by a second order (or parabolic) relationship, i.e.,

$$Y = a_0 + a_1 X + a_2 X^2$$

The third plot suggests a linear model that applies over two different ranges, i.e., it should represent the data

$$Y = a_0 + a_1 X; \quad X_0 < X < X_M$$

and

$$Y = a_0' + a'_1 X; \quad X_M < X < X_L$$

This multiequation model finds application in representing adsorption equilibria, multiparticle size distributions, quantum energy relationships, etc. In any event, a scatter diagram and individual judgment can suggest an appropriate model at an early stage in the analysis.

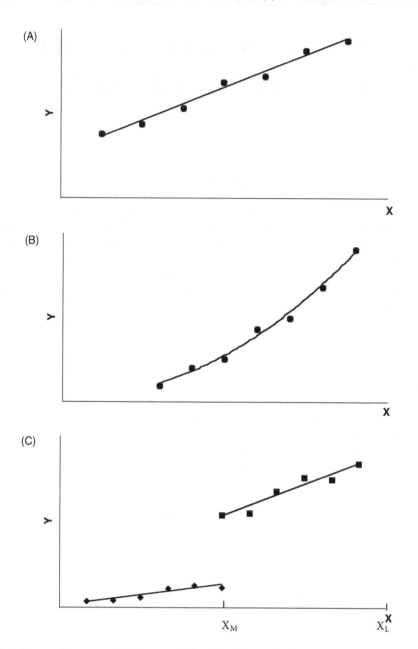

FIGURE 28.1 Scatter diagrams: (A) linear relationship, (B) parabolic relationship, and (C) dual-linear relationship.

Some of the models often employed by technical individuals are as follows:

1.	$Y = a_0 + a_1X$	Linear	(28.1)
2.	$Y = a_0 + a_1X + a_2X^2$	Parabolic	(28.2)
3.	$Y = a_0 + a_1X + a_2X^2 + a_3X^3$	Cubic	(28.3)
4.	$Y = a_0 + a_1X + a_2X^2 + a_3X^3 + a_4X^4$	Quadratic	(28.4)

Procedures to evaluate the regression coefficients a_0, a_1, a_2, etc., are provided as follows. The reader should note that the analysis is based on the method of least squares. This technique provides

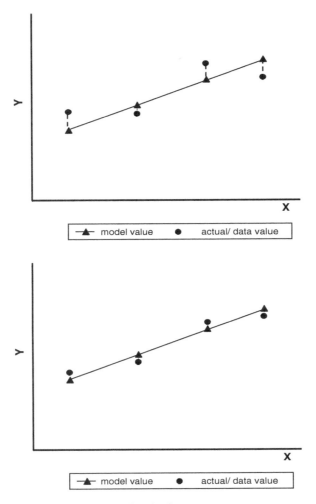

FIGURE 28.2 Error difference: actual and predicted values.

numerical values for the regression coefficients a_i such that the sum of the square of the difference (error) between the actual Y and the Y_e predicted by the equation or model is minimized. This is shown in Figure 28.2.

In Figure 28.2, the dots (experimental value of Y) and triangles (equation or model value of Y, i.e., Y_e) represent the data and model values, respectively. On examining the two figures, one can immediately conclude that the error $(Y - Y_e)$ squared and summed for the four points is less for the lower figure. Also note that a dashed line represents the error. The line that ultimately produces a minimum of the sum of the individual errors squared, i.e., has its smallest possible value, is the regression model (based on the method of least squares). The proof is left as an exercise.

To evaluate a_0 and a_1 for a linear model (see Equation 28.1) one employs the following least-squares algorithm for n data points of Y and X.

$$a_0 n + a_1 \Sigma X = \Sigma Y \tag{28.5}$$

$$a_0 \Sigma X + a_1 \Sigma X^2 = \Sigma XY \tag{28.6}$$

All the quantities given, except a_0 and a_1 can be easily calculated from the data. Because there are two equations and two unknowns, the set of equations can be solved for a_0 and a_1. For this case,

$$a_1 = \frac{n \sum XY - \sum X \sum Y}{n \sum X^2 - \left(\sum X\right)^2} \tag{28.7}$$

Dividing numerator and denominator by n, and defining $\overline{X} = \sum X / n$ and $\overline{Y} = \sum Y / n$, leads to

$$a_1 = \frac{\sum XY - \dfrac{\sum X \sum Y}{n}}{\sum X^2 - \dfrac{\left(\sum X\right)^2}{n}} \tag{28.8}$$

$$= \frac{\sum XY - n\overline{X}\,\overline{Y}}{\sum X^2 - n\overline{X}^2}$$

Using this value of a_1 produces the following equation for a_0:

$$a_0 = \overline{Y} - a_1 \overline{X} \tag{28.9}$$

If the model (or line of regression) is forced to fit through the origin, then the calculated value of $Y_e = 0$ at $X = 0$. For this case, the line of regression takes the form

$$Y_e = a_1 X; \quad a_0 = 0 \tag{28.10}$$

with

$$a_1 = \sum XY \Big/ \sum X^2 \tag{28.11}$$

A cubic model takes the form (Equation 28.3)

$$Y = a_0 + a_1 X + a_2 X^2 + a_3 X^3$$

For n pairs of $X - Y$ values, the constants a_0, a_1, a_2, and a_3 can be obtained by the method of least squares so that $\Sigma (Y - Y_e)^2$ again has the smallest possible value, i.e., is minimized. The coefficients a_0, a_1, a_2, and a_3 are the solution of the following system of four linear equations:

$$a_0 n + a_1 \Sigma X + a_2 \Sigma X^2 + a_3 \Sigma X^3 = \Sigma Y$$

$$a_0 \Sigma X + a_1 \Sigma X^2 + a_2 \Sigma X^3 + a_3 \Sigma X^4 = \Sigma XY$$

$$a_0 \Sigma X^2 + a_1 \Sigma X^3 + a_2 \Sigma X^4 + a_3 \Sigma X^5 = \Sigma X^2 Y \tag{28.12}$$

$$a_0 \Sigma X^3 + a_1 \Sigma X^4 + a_2 \Sigma X^5 + a_3 \Sigma X^6 = \Sigma X^3 Y$$

Because there are four equations and four unknowns, this set of equations can be solved for a_0, a_1, a_2, and a_3. This development can be extended to other regression equations, e.g., exponential, hyperbola, higher order models, etc. Linear multiple regression is considered in Problem REG 8 and Problem REG 9.

The correlation coefficient provides information on how well the model, or line of regression, fits the data. It is denoted by r and is given by

$$r = \frac{\sum XY - \frac{\sum X \sum Y}{n}}{\sqrt{\left[\sum X^2 - \frac{\left(\sum X\right)^2}{n}\right]\left[\sum Y^2 - \frac{\left(\sum Y\right)^2}{n}\right]}} \tag{28.13}$$

or

$$r = \frac{n\sum XY - \sum X \sum Y}{\sqrt{n\left[\sum X^2 - \left(\sum X\right)^2\right]\left[n\sum Y^2 - \left(\sum Y\right)^2\right]}} \tag{28.14}$$

or

$$r = \frac{\sum XY - n\overline{X}\,\overline{Y}}{\sqrt{\left(\sum X^2 - n\overline{X}^2\right)\left(\sum Y^2 - n\overline{Y}^2\right)}} \tag{28.15}$$

This equation can also be shown to take the form

$$r = \pm\left[\frac{\sum\left(\overline{Y} - Y_e\right)^2}{\sum\left(Y - \overline{Y}\right)^2}\right]^{0.5} \tag{28.16}$$

The correlation coefficient satisfies the following six properties:

1. If all points of a scatter diagram lie on a line, then $r = +1$ or -1. In addition, $r^2 = 1$. The square of the correlation coefficient is defined as the coefficient of determination.
2. If no linear relationship exists between the X's and Y's, then $r = 0$. Furthermore, $r^2 = 0$. It can be concluded that r is always between -1 and $+1$, and r^2 is always between 0 and 1.
3. Values of r close to $+1$ or -1 are indicative of a strong linear relationship.
4. Values of r close to 0 are indicative of a weak linear relationship.
5. The correlation coefficient is positive or negative depending on whether the linear relationship has a positive or negative slope. Thus, positive values of r indicate that Y increases as X increases; negative values indicate that Y decreases as X increases.
6. If $r = 0$, it only indicates the lack of linear correlation; X and Y might be strongly correlated by some nonlinear relation, as discussed earlier. Thus, r can only measure the

strength of linear correlations; if the data are nonlinear, one should attempt to linearize before computing r.

Another measure of the model's fit to the data is the standard error of the estimate, or s_e. It is given as

$$s_e = \left[\frac{\sum (Y - Y_e)^2}{n} \right]^{0.5} \tag{28.17}$$

Finally, the equation of the least-squares line of best fit for the given observations on X and Y can be written in terms of the correlation coefficient as follows:

$$Y - \overline{Y} = r \frac{s_y}{s_x} (X - \overline{X}) \tag{28.18}$$

where $\overline{X}, \overline{Y}, s_x, s_y$, and r are the two sample means, the two sample standard deviations, and sample correlation coefficient obtained from the n pairs of observations on X and Y, respectively.

Whether r is significantly different from zero can be checked by testing the hypothesis H_0: $\rho = 0$, where is the population correlation coefficient. Under the assumption that the n pairs of observations on X and Y constitute a random sample from a bivariate normal population, the test statistic

$$\frac{1}{2} \ln \frac{1+r}{1-r} \tag{28.19}$$

is approximately normally distributed with mean 0 and variance $1/(n-3)$ when H_0: $\rho = 0$ is true. The test statistic ultimately leads to

$$Z = \frac{\frac{1}{2} \ln \left(\frac{1+r}{1-r} \right) - 0}{\sqrt{\frac{1}{n-3}}} \tag{28.20}$$

If a given significance level is assumed, e.g., 0.01 or 1%, and if the two tails of this normal distribution are employed as the critical region, a sample value of r is considered significant if it has a value of Z such that it fails outside the region determined from that of the significance level.

It should be noted that the correlation coefficient only provides information on how well the model fits the data. It is emphasized that r provides no information on how good the model is or, to reword this, whether this is the correct or best model to describe the functional relationship of the data.

PROBLEMS AND SOLUTIONS

REG 1: QUALITATIVE EXPLANATIONS

Provide simple qualitative answers to the following two questions:

1. Describe the two main objectives of regression and correlation.
2. Explain the method of least squares in essay form.

Solution

With respect to the first question, one is interested in determining if there is a relationship between two variables (X and Y) and, if such a relationship exists, how it can best be expressed in equation form.

For question 2, the method of least squares is the method of fitting a line to a set of n points in such a way that $\Sigma(Y - Y_e)^2$ has its smallest possible value (is minimized), where the sum is calculated for the given n pairs of values of X and Y.

REG 2: LINEAR REGRESSION OF YIELD OF A BIOLOGICAL REACTION WITH TEMPERATURE

Table 28.1 shows eight pairs of observations on X and Y where Y is the observed percentage yield of a biological reaction at various centigrade temperatures, X. Obtain the least-squares line of regression of Y on X.

Solution

The observed values of Y against the associated values of X are plotted in Figure 28.3.

The scatter diagram appears to exhibit a linear pattern. Equation 28.1 is assumed to apply. The values ΣX, ΣY, ΣXY, ΣX^2, ΣY^2, $(\Sigma X)^2$, $(\Sigma Y)^2$ and n are now calculated:

$$\Sigma X = 1900$$

$$\Sigma Y = 700$$

$$\Sigma XY = 169{,}547.5$$

$$\Sigma X^2 = 477{,}500$$

$$(\Sigma X)^2 = 3{,}610{,}000$$

$$(\Sigma Y)^2 = 490{,}000$$

$$\Sigma Y^2 = 61{,}676.94$$

$$n = 8$$

TABLE 28.1
Temperature — Biological Yield Data

Temperature, °C (X)	% Yield (Y)
150	75.4
175	79.4
200	82.1
225	86.6
250	90.9
275	93.3
300	95.9
325	96.1

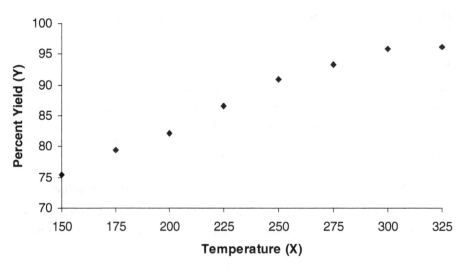

FIGURE 28.3 Scatter diagram of yield vs. temperature for a biological reaction.

The least-squares estimates, a_0 and a_1, are calculated using the equations provided in the Introduction. Applying Equation 28.7,

$$a_1 = \frac{n \sum XY - \sum X \sum Y}{n \sum X^2 - \left(\sum X\right)^2}$$

$$= \frac{(8)(169547.5) - (1900)(700)}{(8)(477500) - (3610000)}$$

$$= 0.126$$

The coefficient is calculated employing Equation 28.9.

$$a_0 = \overline{Y} - a_1 \overline{X}$$

$$= \frac{700}{8} - 0.126\left(\frac{1900}{8}\right)$$

$$= 57.5$$

Thus the describing equation is

$$Y = 57.5 + 0.126\ X; \quad \text{consistent units}$$

The reader should note that if more temperature–yield data become available, thus increasing the number of points, then the calculated line may not be the best representation of all the data; the least-squares solution should then be recomputed using all the data. In addition, the assumed model, e.g., linear, may not be the "best" model.

TABLE 28.2
Reaction Rate Data

Reaction Rate, $-r_A$ (lbmol/ft³-sec)	Concentration, C_A (lbmol/ft³)
48	8
27	6
12	4
3	2

REG 3: CALCULATION OF AVERAGE YIELD VALUES

Use the results of Problem REG 2 to estimate the "average" percentage yield of the biological reaction at 260°C.

Solution

The result of

$$Y = 57.575 + 0.126\ X$$

is employed to calculate the value of Y when $X = 260$. For $X = 260$,

$$Y = 57.575 + 0.126\ (260)$$

$$= 90.3\%$$

Therefore, the estimated average percentage yield at 260°C is 90.3.

REG 4: REACTION RATE ANALYSIS

Pollutant A is undergoing a reaction in a specially controlled laboratory experiment. The data in Table 28.2 have been obtained for the reaction rate, $-r_A$, vs. concentration, C_A. Using the data, estimate the coefficient k_A and α in the equation below:

$$-r_A = k_A C_A{}^\alpha$$

Solution

As discussed in the Introduction, given experimental data for Y measured at known values of X, an often encountered problem is to identify the functional relationship between the two. One invariably starts by constructing an X–Y plot. The goal is to find a simple relation between X and Y, e.g., to seek a straight line. If the X–Y plot is not straight, one should try to straighten it by replotting, using other functions, other scales or both.

Some functions readily linearize; examples include the exponential

$$Y = Ae^{mX}$$

and the power law

$$Y = X^{\alpha}$$

These linearize by taking the natural logarithms:

$$\ln Y = \ln A + mX; \quad \text{exponential}$$

and

$$\ln Y = \alpha \ln X; \quad \text{power law}$$

Thus, exponentials yield straight lines on semilog plots, whereas power laws yield straight lines on log–log plots.

For the problem at hand, linearize the equation by taking the natural logarithm (ln) of both sides of the equation.

$$-r_A = k_A C_A{}^{\alpha}$$

$$\ln(-r_A) = \ln(k_A) + \alpha \ln(C_A)$$

Change variables to Y and X, so that

$$Y = A + BX$$

Regress the preceding four data points using the method of least squares, where $A = \ln (k_A)$ and $B = \alpha$. Once again, the method of least squares requires that the sum of the errors squared between the data and the model is minimized.

$$\ln(3) = A + B \ln(2)$$

$$\ln(12) = A + B \ln(4)$$

$$\ln(27) = A + B \ln(6)$$

$$\ln(48) = A + B \ln(8)$$

Employ Equation 28.7 and Equation 28.9. The linear equation coefficients A and B are given by

$$A = -0.2878$$

$$B = 2.0$$

These may be obtained through longhand calculation. However, they are more often obtained with the aid of computer software.

Take the inverse natural logarithm of A to obtain k_A:

$$A = \ln k_A = 0.75$$

$$B = 2.0$$

The equation for the rate of reaction is therefore

$$-r_A = 0.75 C_A{}^{2.0}$$

TABLE 28.3
Transistor Resistance — Failure Time Data

Resistance (X)	Failure Time (Y)	Resistance (X)	Failure Time (Y)
43	32	36	31
36	36	44	37
29	20	29	24
44	45	39	46
32	34	46	43
48	47	42	33
35	29	30	25
30	25	30	25
33	32	35	35
46	47	45	46

REG 5: CORRELATION OF RESISTANCE AND FAILURE TIME OF RESISTORS

The resistance in ohms and failure time in minutes of a random sample of 20 transistors were recorded as shown in Table 28.3.

Calculate the linear correlation coefficient, r.

Solution

Calculate ΣX, ΣY, ΣXY, ΣX^2, ΣY^2, and n for the assumed linear model.

$$\Sigma X = 752$$

$$\Sigma Y = 692$$

$$\Sigma XY = 26{,}952$$

$$\Sigma X^2 = 29{,}104$$

$$\Sigma Y^2 = 25{,}360$$

$$n = 20$$

Obtain r using Equation 28.14:

$$r = \frac{n\sum XY - \sum X \sum Y}{\sqrt{\left(n\sum X^2 - \left(\sum X\right)^2\right)\left(n\sum Y^2 - \left(\sum Y\right)^2\right)}}$$

$$= \frac{20(26952) - 752(692)}{\sqrt{\left(20(29104) - 752^2\right)\left(20(25360) - 692^2\right)}}$$

$$= 0.86$$

REG 6: Significance Test for the Correlation Coefficient

Determine whether the observed value of r (calculated in Problem REG 5) is significantly different from zero at the 5% level of significance.

Solution

Test H_0: $\rho = 0$ against H_1: $\rho \neq 0$ with $\alpha = 0.05$. From Equation 28.19, when H_0 is true, the term

$$\frac{\frac{1}{2}\ln\frac{1+r}{1-r}}{\sqrt{\frac{1}{n-3}}}; \quad \sigma^2 = \frac{1}{n-3}$$

is a standard normal variable

$$Z = \frac{1}{2}\frac{\left[\ln\frac{1+r}{1-r}\right]}{\sqrt{\left[\frac{1}{n-3}\right]}}$$

Solving for Z gives

$$Z = \frac{1}{2}\frac{\left[\ln\frac{1.85}{0.15}\right]}{\sqrt{\frac{1}{20-3}}}$$

$$= 5.33$$

From the table of the standard normal distribution, $P\left(|Z| > 1.96\right) = 0.05$. Therefore, H_0: $\rho = 0$ is rejected in favor of H_1: $\rho \neq 0$ because the observed value of Z lies outside the interval $(-1.96, 1.96)$ if $\alpha = 0.05$. This result may be interpreted as follows: because the observed value of Z is 5.02, reject H_0: $\rho = 0$ and conclude that the observed value of r, 0.85, is significantly different from zero at the 5% level of significance.

REG 7: Least-Squares Line of Best Fit

Using the results of Problem REG 5, determine the least-squares line of best fit for the data provided in Problem REG 5.

Solution

To obtain the equation of the least-squares line of best fit, substitute into Equation 28.18,

$$Y - \overline{Y} = r\frac{s_y}{s_x}(X - \overline{X})$$

where

$$\frac{s_y}{s_x} = \sqrt{\frac{n\left(\sum Y^2\right) - \left(\sum Y\right)^2}{n\left(\sum X^2\right) - \left(\sum X\right)^2}}$$

This yields

$$Y - \frac{692}{20} = 0.86\sqrt{\frac{20(25360) - 692^2}{20(29104) - 752^2}}\left(X - \frac{752}{20}\right)$$

$$Y - 34.6 = 0.86(1.31)(X - 37.6)$$

which upon simplification becomes

$$Y = 1.13X - 7.76$$

This represents the least-squares line of best fit for the given sample of observations on X and Y. The terms s_y^2 and s_x^2 can now be calculated.

$$s_y^2 = 20(25360) - 692^2$$

$$= 28336$$

$$s_x^2 = 20(29104) - 752^2$$

$$= 16576$$

REG 8: Multiple Regression Procedure

Provide the algorithm for the regression of two independent variables.

Solution

Consider a variable Z that is dependent upon two independent variables X_1 and X_2. If the relationship between Y to X_1 and X_2 is assumed to be linear, i.e.,

$$Y = a_0 + a_1 X_1 + a_2 X_2 \tag{28.21}$$

it can be shown that a_0, a_1, and a_2 must be the solution of the following system of three (linear) equations:

$$a_0 n + a_1 \Sigma X_1 + a_2 \Sigma X_2 = \Sigma Y$$

$$a_0 \Sigma X_1 + a_1 \Sigma X_1^2 + a_2 \Sigma X_1 X_2 = \Sigma X_1 Y \tag{28.22}$$

$$a_0 \Sigma X_2 + a_1 \Sigma X_1 X_2 + a_2 \Sigma X_2^2 = \Sigma X_2 Y$$

TABLE 28.4
PCB Data for Various Incinerator Lengths and Diameters

Incinerator Number	PCB Emissions, E (ppm)	Incinerator Length, L (ft)	Incinerator Diameter, D (ft)
1	24	42	14
2	29	45	16
3	27	44	15
4	27	43	15
5	25	45	13
6	26	43	14
7	28	46	14
8	30	44	16
9	28	45	16
10	28	44	15

The aforementioned development can be extended to other regression equations, e.g., exponential, hyperbola, and other higher-order models.

REG 9: MULTIPLE REGRESSION ANALYSIS

The United States Environmental Protection Agency (USEPA) conducted a study that yielded information on PCB emissions from ten rotary kiln hazardous waste incinerators. The data took the form of the emissions (ppm) as a function of kiln diameter and kiln length.

Perform a least-squares linear regression analysis to obtain an equation that predicts emissions (E) in terms of the kiln length (L) and diameter (D). Assume the describing equation is given by

$$E = a_0 + a_1 L + a_2 D$$

where a_0, a_1, and a_2 are the regression coefficients.

Solution

From Equation 28.22, the equations to be solved for a_0, a_1, and a_2 are

$$na_0 + a_1 \Sigma L + a_2 \Sigma D = \Sigma E$$

$$a_0 \Sigma L + a_1 \Sigma L^2 + a_2 \Sigma LD = \Sigma EL$$

$$a_0 \Sigma D + a_1 \Sigma LD + a_2 \Sigma D^2 = \Sigma ED$$

The following values are then calculated:

$$\Sigma E = 272$$

$$\Sigma L = 441$$

$$\Sigma D = 147$$

$$\Sigma EL = 12,005$$

$$\Sigma ED = 4,013$$

$$\Sigma LD = 6,485$$

$$\Sigma L^2 = 19,416$$

$$\Sigma D^2 = 2,173$$

Substituting into the preceding trio of equations yields

$$272 = 10a_0 + 441a_1 + 147a_2$$

$$12005 = 441a_0 + 19461a_1 + 6485a_2$$

$$4013 = 147a_0 + 6485a_1 + 2173a_2$$

Solving simultaneously gives

$$a_0 = 13.828$$

$$a_1 = 0.564$$

$$a_2 = 1.099$$

The describing equation is therefore

$$E = 13.828 + (0.564)L + (1.099)D$$

REG 10: CORRELATION COEFFICIENT SIGNIFICANCE

A correlation coefficient based on a sample size of 28 was computed to be 0.220. Can one conclude that the corresponding population coefficient differs from zero? Perform the test at a significance level of

1. 0.10
2. 0.05

Solution

The test hypotheses are

$$H_0: \rho = 0$$

$$H_1: \rho \neq 0$$

For this calculation, the critical value of t can be shown to given by

$$t = \frac{r\sqrt{n-2}}{\sqrt{1-r^2}} \tag{28.23}$$

Substituting gives

$$= \frac{(0.220)(26)^{1/2}}{\sqrt{1-(0.220)^2}}$$

$$= 1.15$$

For $\nu = 26$ (case 1) refer to Table A.2 (Appendix A) and apply a two-sided test:

$$t_{0.050} = 1.706$$

For $\nu = 26$ (case 2) apply a two-sided test:

$$t_{0.025} = 2.056$$

Since $t > t_{0.025} > t_{0.050}$, one cannot reject H_0 at either significance level.

REG 11: REGRESSION COEFFICIENT CONFIDENCE INTERVAL

A regression analysis of nanoparticle emissions (E) as a function of the process operating temperature (T) produced the following result. For the model

$$E = a_0 + a_1 T$$

$$E = 9.8738 + 0.042834T$$

Peripheral calculation on the eight data points (with E and T replaced by Y and X, respectively) were also provided:

$$s_{YX} = 4.3582$$

$$s_X = 244.94$$

$$\overline{X} = 5200/8$$

Calculate the 95% confidence interval for a_1.

Solution

Once again, for this two-sided test with $\alpha/2 = 0.025$ and $n = 6$ (from Table A.2 (Appendix A)),

$$t_{0.025;6} = \pm 2.45$$

The interval for a_1, denoted as a_1', can be shown to be given by

$$a_1 + (t_{\alpha/2})\frac{s_{YX}}{s_X\sqrt{n-1}} < a_1' < a_1 + (t_{1-(\alpha/2)})\frac{s_{YX}}{s_X\sqrt{n-1}} \tag{28.24}$$

Because

$$t_{\alpha/2}\frac{S_{YX}}{S_X\sqrt{n-1}}=\frac{(2.45)(4.3582)}{(244.94)(\sqrt{7})}$$

$$=0.001648$$

the interval for a_1 is therefore (with $t_{\alpha/2}=t_{1-(\alpha/2)}$),

$$0.042834\text{-}0.001648 < a_1' < 0.0042834 + 0.001648$$

$$0.041186 < a_1' < 0.044482$$

The reader should note that various forms of Equation 28.24 have appeared in the literature.

REG 12: CORRELATION COEFFICIENT CONFIDENCE INTERVAL

Refer to Problem REG 11. Obtain the 95% confidence interval for the regression coefficient a_0.

Solution

The interval estimate for a_0, again denoted as a_0' is

$$a_0+(t_{\alpha/2})s_{YX}\sqrt{\frac{1}{n}+\frac{(\overline{X})^2}{(n-1)(S_X)^2}} < a_0' < a_0+(t_{1-(\alpha/2)})s_{YX}\sqrt{\frac{1}{n}+\frac{(\overline{X})^2}{(n-1)(s_X)^2}}$$

Note that

$$t_{(\alpha/2)}s_{YX}\sqrt{\frac{1}{n}+\frac{(\overline{X})^2}{(n-1)(S_X)^2}}=(2.45)(4.3582)\sqrt{\frac{1}{8}+\frac{(650)^2}{(7)(244.94)^2}}$$

$$=10.838$$

Therefore, the interval estimate is (with $t_{\alpha/2}=t_{1-(\alpha/2)}$ once again)

$$9.8738 - 10.838 < a_0' < 9.8738 + 10.838$$

$$-0.964 < a_0' < 20.712$$

Furthermore, because this interval encompasses zero, one can conclude that a_0 is not significantly different from zero.

29 NPT. Nonparametric Tests

INTRODUCTION

Classical tests of statistical hypotheses are often based on the assumption that the populations sampled are normal. In addition, these tests are frequently confined to statements about a finite number of unknown parameters on which the specification of the probability distribution function (pdf) of the random variable under consideration depends. Efforts to eliminate the necessity of restrictive assumptions about the population sampled have resulted in statistical methods called *nonparametric*, directing attention to the fact that these methods are not limited to inferences about population parameters. These methods are also referred to as *distribution free*, emphasizing their applicability in cases in which little is known about the functional form of the pdf of the random variable observed. Nonparametric tests of statistical hypotheses are tests whose validity generally requires only the assumption of continuity of the cumulative distribution function (cdf) of the random variable involved.

The sign test for paired comparison provides a good illustration of the nonparametric approach. In the case of paired comparisons, the data consist of pairs of observations on a random variable. Examples are observations on the number of chemical plant accidents before and after the institution of a safety program, observations on air pollution measured by two different instruments, and observations on the weights of subjects before and after a weight reduction program. Interest centers on whether or not the paired observations "differ significantly."

A t test using Student's distribution assumes the difference of paired observations to be independently and normally distributed with a common variance; the hypothesis tested is that the mean of the normal population of differences is equal to zero. In the sign test no assumption is required concerning the form of the distribution of the differences; the hypothesis tested is that each of the differences has a probability distribution with median equal to zero, and, therefore, the probability of a plus sign difference as opposed to a negative sign difference is equal to 0.5. Interestingly, the probability distribution need not be the same for all differences. The test statistic becomes the number of plus-signed differences. When the null hypothesis is true, the test statistic has a binomial distribution with P equal to 0.5, and n equal to the number of pairs of observations. This test is detailed in Problem NPT 2 to Problem NPT 5.

The reader should note that the sign test makes use only of the sign of the difference — not the magnitude. This inefficiency is compensated for, however, by the aforementioned freedom from restrictive assumptions concerning the probability distribution of the sample observations, by the adaptability of many nonparametric tests to situations in which the observations are not susceptible to quantitative measurement, and by usually increased computational simplicity, especially in the case of small samples.

The reader should also note that when ties occur in ranking, the mean value of the ranking numbers is applied to each of the tied observations. For example, if two observations are tied for the 6th rank, they are both ranked as 6.5 and the next observation is rank number 8. If three observations are tied for 11th rank, they are all ranked 12, and the next observation after the tie is rank 14.

A large group of nonparametric tests makes use of the method of replacing observations in a combined sample by their ranks after these observations have been arranged in ascending order of magnitude. Rank 1 is assigned to the smallest (numerical) observation, rank 2 to the next, and so on.

The Wilcoxon test of the hypothesis that two samples come from the same population is based on the sum of ranks on one sample after the combined sample has been ranked as already indicated. The sum of the ranks of the combined sample of n observations is $n(n+1)/2$. For example, consider the ranked observations

$$1, 2, 3, 4, 5$$

Because $n = 5$, the sum, S, is

$$S = \frac{n(n+1)}{2} = \frac{5(5+1)}{2}$$
$$= 15$$

This is exactly given by

$$S = 1+2+3+4+5$$
$$= 15$$

The sum of the ranks in one sample then determines (by difference) the sum of ranks in the other. Significantly small or significantly large values of either sum of ranks can lead to rejection of the hypothesis that the two samples come from the same population.

PROBLEMS AND SOLUTIONS

NPT 1: SIGN TEST INEFFICIENCY

Describe some of the features and inefficiencies associated with the sign test.

Solution

The sign test shares with other nonparametric tests a certain inefficiency in the sense that they do not make use of all the information contained in the sample. Whereas the t test makes use of the actual magnitude of the difference of paired observations, the sign test makes use of only the sign of the difference. The magnitude of the difference need not be specified. Thus, sign tests find application with data/information that can be classified as +/−, on/off, 0/1, yes/no, etc. As noted earlier, the aforementioned inefficiency is compensated for, however, by freedom from restrictive assumptions concerning the probability distribution of many nonparametric tests to situations in which the observations are not susceptible to quantitative measurement.

NPT 2: NONPARAMETRIC TEST RANKING EXAMPLES

Provide examples of ranking tests.

Solution

As noted in the Introduction, ranking refers to the numbering of observations in an order based on qualitative or quantitative bases or criteria. Examples include the following:

1. Listing chemicals in order of health effects
2. Listing chemicals in order of hazard effects
3. Judging the best 3-year-old as the Kentucky Derby approaches
4. Ranking the quality of service at various casinos

Note that the n "observations" are ranked from 1 to n with reference to the classification of interest. The numbers indicate only the order in the classification and do not necessarily relate to any quantitative measure. Thus, ranking is used when there is no specific quantitative measure available. Ranking cities according to their distances from New York City could give information more applicable to a particular situation than if the actual distances were tabulated.

Ranking may also be used to determine whether two different classifications of the same observations may be related without requiring an actual quantitative relation. For example, with reference to (4) above, one might rank the casinos according to annual profit in order to ascertain if there is any relationship between service and profit.

NPT 3: SIGN TEST ALGORITHM

Outline how to perform a one-tailed and a two-tailed sign test for paired comparisons.

Solution

In a typical calculation, the signs of the differences are noted, and number of positive (preferable to negative) signs are assigned a statistic, say X. For example, if $X_1, X_2, X_3, \ldots, X_n$ represent values from a sample (from a continuous probability distribution), one could test the following one-sided (right tail) hypothesis problem:

$$H_0: \mu = \mu_0$$

$$H_1: \mu > \mu_0$$

For H_0, X has a binomial distribution $(n, P = 1/2)$. For the alternative hypothesis H_1, X takes on a larger value with a binomial distribution (n, P), where P is the probability that X is to the right of μ_0. If α is the tolerated probability of a Type I error, H_0 is rejected if $X \geq X_\alpha$, where X is selected according to the binomial distribution $(n, 1/2)$ such that $P(X \geq X_\alpha)$ is as large as possible although not exceeding. In effect, if $P(X \geq X_\alpha)$ is less than α, reject H_0.

For the two-sided case where

$$H_1: \mu \neq \mu_0$$

one would reject H_0 if X_α is either too small or too large, using $\alpha/2$ at most, for each trial.

NPT 4: DIGESTER/REACTOR SIGN TEST COMPARISON

The following eight temperature readings (°F) were recorded in a newly designed digester/reactor:

$$530, 590, 570, 560, 490, 510, 540, 610$$

Perform a two-sided test for the following hypotheses using 0.01 as the tolerated probability of a Type I error:

$$H_0: \mu = 500°F$$

$$H_1: \mu \neq 500°F$$

Solution

Let X represent the number of positive differences from 500°F. For a binomial distribution with $n = 8$ and $P = 0.5$,

$$P(X=0) = \frac{8!(0.5)^8}{8!(8-8)!} = (0.5)^8$$

$$= \frac{1}{256}$$

$$= 0.003906$$

$$= 0.3906\%$$

and similarly,

$$P(X=8) = \frac{1}{256}$$

$$= 0.003906$$

$$= 0.3906\%$$

Combining both results gives

$$P(X = 0) + P(X = 8) = 0.003906 + 0.003906$$

$$= 0.00781 \leq 0.01$$

Thus, the hypothesis H_0 is rejected if $X = 0$ or $X = 8$. Alternatively, one could also note that

$$P(X = 0) = P(X = 8) = 0.003906 \leq 0.005$$

This also indicates a failure to reject H_0 at the 1% level for $X = 0$ or $X = 8$. Because the number of + signs among the $(X_i - 500)$ data/calculations is 7, and for the probabilities for $X \neq 0$ and $X \neq 8$, the tolerated probability exceeds this value so that H_0 is accepted, i.e., the hypothesis that the median temperature is 500°F is accepted.

NPT 5: REACTOR TEST COMPARISON AT THE 0.005 TYPE I ERROR

Refer to Problem NPT 4. Resolve the problem using 0.005 as the tolerated probability of a Type I error.

Solution

For this error,

$$0.00781 \geq 0.005$$

or

$$0.00391 \geq 0.0025$$

This indicates a rejection of H_0, i.e., the hypothesis that the median temperature is 500°F is rejected.

TABLE 29.1
Sulfur Dioxide (SO$_2$) Data for
Two Analyzers

	SO$_2$ Concentration (ppm)	
Day	Instrument A	Instrument B
1	3	1
2	7	5
3	16	14
4	14	17
5	20	18
6	11	10
7	9	6
8	12	11
9	20	19
10	23	21
11	14	16
12	8	6
13	4	5
14	10	8
15	12	9

NPT 6: Paired Comparison of Instruments Measuring Sulfur Dioxide Concentrations

As part of an air pollution experiment, two instruments for measuring sulfur dioxide concentration (ppm) in the atmosphere are compared over a 15-d period. Daily readings obtained by both instruments over a 15-d period are provided in Table 29.1.

Test the hypothesis that the difference in paired instrument readings is not significant against the alternative hypothesis that instrument A readings are higher than instrument B readings. Use 0.05 as the tolerated probability of a Type I error.

Solution

Record the sign of the difference (A − B) between the observations in each pair (see Table 29.2). X denotes the number of plus sign differences obtained, 12. Therefore, $X = 12$. Assuming H_0 is

TABLE 29.2
Sign Difference for Two Sulfur Dioxide (SO$_2$)
Analyzers

Pair	Sign of Difference	Pair	Sign of Difference
(3, 1)	+	(20, 19)	+
(7, 5)	+	(23, 21)	+
(16, 14)	+	(14, 16)	−
(14, 17)	−	(8, 6)	+
(20, 18)	+	(4, 5)	−
(11, 10)	+	(10, 8)	+
(9, 6)	+	(12, 9)	+
(12, 11)	+		

true, calculate $P(X \geq x)$ where x is the value just obtained. Apply the binomial distribution to the sign test

$$P(X \geq 12) = \sum_{x=12}^{15} \frac{15!}{x!(15-x)!}(0.5)^{15}$$

$$= 0.0139 + 0.0032 + 0.0005 + 0.0000$$

$$= 0.0176$$

Compare the probability obtained with 0.05. Because the P-value, 0.0176, is less than 0.05, reject H_0. Thus, instrument A readings of sulfur dioxide concentration tend to be higher than instrument B readings.

NPT 7: PAIRED TEST: DIFFERENT TYPE I ERROR

Refer to Problem NPT 6. Resolve the problem if the tolerated probability of a Type I error is 0.01.

Solution

Because the P-value remains the same, i.e., 0.0176, the probability is now higher than 0.01. Therefore, accept H_0.

NPT 8: THE WILCOXON TEST

Provide calculational details of the Wilcoxon test.

Solution

Suppose one sample consists of m observations and the other sample consists of n observations. Let R_1 denote the sum of the ranks assigned to the m observations in one sample after the combined sample of $N = m + n$ observations has been ranked from 1 to N. For large sample sizes, one can show that R_1 is approximately normally distributed with

$$\mu = m(m + n + 1)/2 \tag{29.1}$$

and

$$\sigma = mn(m + n + 1)/12 \tag{29.2}$$

Therefore, the test statistic

$$W = \frac{R_1 - \dfrac{m(m+n+1)}{2}}{\sqrt{\dfrac{mn(m+n+1)}{12}}} \tag{29.3}$$

is approximately distributed as a standard normal variable. Good approximations are obtained when both m and n are greater than 10. A numerical application follows in Problem NPT 9.

TABLE 29.3
Ranking Procedure

Observation	Rank	Observation	Rank
37.5	1	44.0	15[a]
38.0	2	44.1	16[a]
39.0	3.5[a]	44.2	17[a]
39.0	3.5	44.4	18[a]
39.5	5	44.5	19[a]
39.6	6	44.7	20[a]
40.0	7	44.8	21[a]
41.5	8	47.0	22[a]
42.0	9[a]	47.1	23
42.5	10.5[a]	47.2	24
42.5	10.5[a]	47.6	25
42.7	12[a]	47.8	26
42.8	13[a]	48.5	27
43.0	14[a]	49.8	28

[a] Ranks of the 15 observations on Brand I.

NPT 9: Comparison of Disconnect Forces Required to Separate Electrical Connectors

Electrical connectors in the missile industry are electromechanical components used to connect and disconnect electrical circuits. The force required to separate electrical connectors is critical because it is frequently involved in the separation of moving vehicles. The following data represent samples of observations on the disconnect force in pounds (force) for two brands of connectors, I and II.

Brand I: 44.2, 44.5, 39.0, 42.5, 43.0, 44.8, 42.0, 44.1, 42.7, 42.8, 44.0, 44.7, 44.4, 47.0, 42.5
Brand II: 47.8, 47.2, 41.5, 40.0, 38.0, 39.6, 47.6, 48.5, 37.5, 39.5, 49.8, 47.1, 39.0

Test the hypothesis that the two samples come from the same population. Use 0.05 as the tolerated probability of a Type I error.

Solution

Identify the sample sizes m and n:

The number of observations on Brand I connector (m) = 15
The number of observations on Brand II connector (n) = 13

The observations in the combined sample of $m + n$ observations are ranked. The ranks of the observations in the first two samples (I and II) are identified in Table 29.3.

Evaluate R_1, the sum of the ranks of the m observations in the first sample. This represents the sum of the ranks of the observations on Brand I. This represents

$$R_1 = 3.5 + 0 + 10.5 + \ldots + 21 + 22$$

$$= 220.5$$

The test statistic for this application is W. The term W is defined in Equation 29.3.

$$W = \frac{R_1 - \dfrac{m(m+n+1)}{2}}{\sqrt{\dfrac{mn(m+n+1)}{12}}}$$

Substitution yields

$$W = \frac{220.5 - \dfrac{15(15+13+1)}{2}}{\sqrt{\dfrac{(15)(13)(15+13+1)}{12}}}$$

$$= 0.138$$

Hypothesis H_0 is rejected if the value of W is greater than $Z_{\alpha/2}$ or is less than $-Z_{\alpha/2}$ (a two-tailed test) where $Z_{\alpha/2}$ is the tabulated value that a standard normal variable exceeds with probability $\alpha/2$. For this case,

$$Z_{\alpha/2} = Z_{0.025} = 1.96$$

Because the observed value of W lies between -1.96 and 1.96, on would fail to reject the null hypothesis that the two samples come from the same population. Thus, there is insufficient evidence to conclude that Brand I and Brand II electrical connectors differ significantly with respect to disconnect force required to separate electrical connectors.

NPT 10: Comparison Analysis: Different Type I Error

Refer to Problem NPT 9. Resolve the problem if the tolerated probability of a Type I error is 0.10.

Solution

The test statistic W remains the same, i.e., $W = 0.138$. The Z value needs to be obtained for $\alpha = 0.10$, or $\alpha/2 = 0.05$. This value is

$$Z_{\alpha/2} = Z_{0.05} = 1.645$$

Because W still lies between -1.645 and $+1.645$, one would fail to reject the null hypothesis once again.

Note that a large positive value of W indicates that Brand I observations occur disproportionately in the higher ranks of the combined sample. Large negative values of W indicate that Brand I observations occupy a disproportionate share of the lower ranks of the combined sample.

30 ANV. Analysis of Variance

INTRODUCTION

Analysis of variance is a statistical technique featuring the splitting of a measure of total variation into components measuring variation attributable to one or more factors or combinations of factors. The simplest application of analysis of variance involves data classified in categories (levels) of one factor.

Suppose, for example, that k different food supplements for cows are to be compared with respect to milk yield. Let X_{ij} denote the milk yield of the jth cow in a random sample of n_i cows receiving the ith food supplement; $i = 1, ..., k$. Let n, X_i, and X denote, respectively, the total sample size, the mean of the observations in the ith sample, and the mean of the observations in all k samples:

$$n = \sum_{i=1}^{k} m_i = km \tag{30.1}$$

$$\overline{X}_{i.} = \frac{\sum_{j=1}^{m_i} X_{ij}}{m_i} \tag{30.2}$$

$$\overline{X} = \frac{\sum_{i=1}^{k} \sum_{j=1}^{m_i} X_{ij}}{n} = \frac{\sum_{i=1}^{k} m_i \overline{X}_{i.}}{n} \tag{30.3}$$

The term $\sum_{i=1}^{k} \sum_{j=1}^{m_i} (X_{ij} - \overline{X})^2$, a measure of the total variation of the observations, can be algebraically split into components (derivation not provided) as follows:

$$\sum_{i=1}^{k} \sum_{j=1}^{m_i} (X_{ij} - \overline{X})^2 = \sum_{i=1}^{k} \sum_{j=1}^{m_i} [(X_{ij} - \overline{X}_{i.}) + (\overline{X}_{i.} - \overline{X})]^2$$

$$= \sum_{i=1}^{k} \sum_{j=1}^{m_i} (X_{ij} - \overline{X}_{i.})^2 + \sum_{i=1}^{k} m_i (\overline{X}_{i.} - \overline{X})^2 \tag{30.4}$$

Or, in essay form,

$$\begin{array}{c} \text{Total sum of squares (TSS)} = \text{Within group sum} + \text{Among group sum of} \\ \text{of squares (RSS)} \qquad \text{squares (GSS)} \end{array} \tag{30.5}$$

TABLE 30.1
Summary

Group (i)/Row (j)	$i = 1$	$i = 2$	$i = i$	$i = k$
$j = 1$	X_{11}	X_{12}	X_{1i}	X_{1k}
$j = 2$	X_{21}	X_{22}	X_{2i}	X_{2k}
$j = j$	X_{j1}	X_{j2}	X_{ji}	X_{jk}
$j = m$	X_{m1}	X_{m2}	X_{mi}	X_{mk}
	$T_{1.}$	$T_{2.}$	T_i	$T_{k.}$
	M_1	M_2	M_i	M_k

Note: $T_{1.} = X_{11} + X_{21} + ... + X_{j1} + ... + X_{m1}$; $T_{2.} = X_{12} + X_{22} + ... + X_{j2} + ... + X_{m2}$; $T_{k.} = X_{1k} + X_{2k} + ... + X_{jk} + ... + X_{mk}$; $T_{..} = T_{1.} + T_{2.} + ... + T_{i.} + ... + T_{k.}$; $M_1 = T_{1.}/m$; $M_2 = T_{2.}/m$; $M_i = T_{i.}/m$; $M_k = T_{k.}/m$; and $n = mk$

Thus, the measure of total variation has been expressed in terms of two components, the first measuring variation within the k samples, and the second measuring the variation among the k samples. In the terminology of analysis of variance, the within group sum of squares is also called the *residual sum of squares* (RSS) or the *error sum of squares*. The analysis of variance statistic for testing the null hypothesis of no significant difference among the food supplements with respect to milk yield features comparison of the among group sum of squares with the within group sum of squares. If the mean among group sum of squares, i.e., the among group sum of squares divided by $k - 1$ (the degrees of freedom), is large relative to the mean within group sum of squares, i.e., the within group sum of squares divided by $n - k$, there is evidence for contradicting the null hypothesis.

The among group of squares (GSS) and the residual sum of squares (RSS) may also be written as

$$GSS = \sum_{i=1}^{k} (T_{i.}^2 / m_i) - (T_{..}^2 / n) \tag{30.6}$$

$$RSS = \sum_{j=1}^{m_i} \sum_{i=1}^{k} (X_{ij})^2 - \sum_{i=1}^{k} (T_{i.}^2 / m_i) \tag{30.7}$$

where $T_{i.}$ is the total of the js in group i; $j = 1, 2, ..., m_i$, $T_{..}$ is the grand total, and n is equal to km. The total sum of squares (TSS) is given by the sum of the results of Equation 30.6 and Equation 30.7. A summary of the analysis is provided in Table 30.1.

The precise form of the analysis of variance test statistic (ANVT) is

$$ANVT = \frac{n-k}{k-1} \frac{\sum_{i=1}^{k} m_i (\overline{X}_{i.} + \overline{X})^2}{\sum_{i=1}^{k} \sum_{j=1}^{m_i} (X_{ij} - \overline{X}_{i.})^2} \tag{30.8}$$

Under the null hypothesis, this statistic has an F distribution with $k - 1$ and $n - k$ degrees of freedom. A random variable having such a distribution is indicated by $F_{k-1,n-k}$. When the observed value of the test statistic exceeds $F_{k-1,n-k,\alpha}$ where $P(F_{k-1,n-k} > F_{k-1,n-k,\alpha}) = \alpha$, the tolerated probability of a Type I error, the null hypothesis is rejected.

One may also calculate the following two mean sums of squares. For the among group (M),

$$MGSS = \frac{GSS}{k-1} \tag{30.9}$$

$$MRSS = \frac{RSS}{n-k} \tag{30.10}$$

where $k-1$ is the group degree of freedom and $n-k$ is the residual degree of freedom.

The analysis of variance test statistic, ANVT is

$$ANVT = \frac{MGSS}{MRSS} \tag{30.11}$$

This is a modified form of Equation 30.8.

Another simple application of the analysis of variance to the study of variation attributable to two factors is the two-way classification in which rc observations are classified in r rows and c columns with one observation in each cell. The rows represent the levels of one factor and the columns, the levels of the other factor. Let $X_{ij}, i = 1, \ldots, r; j = 1, \ldots, c$ be the observation in the ith row and jth column. The term $X_{i.}, X_{.j}$; and X will denote, respectively, the mean of the observations in the ith row, the mean of the observations in the jth column, and the mean of all the observations. Accordingly,

$$\overline{X}_{i.} = \frac{\sum_{j=1}^{c} X_{ij}}{c} \tag{30.12}$$

$$\overline{X}_{.j} = \frac{\sum_{i=1}^{r} X_{ij}}{r} \tag{30.13}$$

$$\overline{X} = \frac{\sum_{i=1}^{r}\sum_{j=1}^{c} X_{ij}}{rc} \tag{30.14}$$

It can be shown that the total sum of squares can be partitioned as follows:

$$\sum_{i=1}^{r}\sum_{j=1}^{c}(X_{ij} - \overline{X})^2 = \sum_{i=1}^{r}\sum_{j=1}^{c}[(\overline{X}_{i.} - \overline{X}) + (\overline{X}_{.j} - \overline{X}) + (X_{ij} - \overline{X}_{i.} - \overline{X}_{.j} + \overline{X})]^2$$

$$= c\sum_{i=1}^{r}(\overline{X}_{i.} - \overline{X})^2 + r\sum_{j=1}^{c}(\overline{X}_{.j} - \overline{X})^2 + \sum_{i=1}^{r}\sum_{j=1}^{c}(X_{ij} - \overline{X}_{i.} - \overline{X}_{.j} + \overline{X})^2 \tag{30.15}$$

The three components into which the total sum of squares (TSS2) is partitioned are designated as the row sum of squares (RSS2), the column sum of squares (CSS2), and the residual sum of squares (RRSS2), in that order. These four terms from Equation 30.15 may be rewritten as

$$RSS2 = c\sum_{i=1}^{r}(\overline{X}_{i.} - \overline{X})^2 = \frac{\sum T_{i.}^2}{c} - \frac{T_{..}^2}{rc} \tag{30.16}$$

$$CSS2 = r\sum_{j=1}^{c}(\overline{X}_{.j} - \overline{X})^2 = \frac{\sum T_{.j}^2}{r} - \frac{T_{..}^2}{rc} \tag{30.17}$$

$$TSS2 = \sum_{i=1}^{r}\sum_{j=1}^{c}(X_{ij} - \overline{X})^2 = \sum_{i=1}^{r}\sum_{j=1}^{c}X_{ij}^2 - \frac{T_{..}^2}{rc} \tag{30.18}$$

The residual sum of squares can be obtained by subtracting the sum of the row and column sums of squares from the total sum of squares. Therefore, the residual sum of squares is given as

$$RRSS2 = \sum_{i=1}^{r}\sum_{j=1}^{c}X_{ij}^2 - \sum\frac{T_{i.}^2}{c} - \sum\frac{T_{.j}^2}{r} + \frac{T_{..}}{rc} \tag{30.19}$$

The mean row sum of squares and the mean column sum of squares are obtained by dividing the sum of squares by $r - 1$ and $c - 1$, respectively. The mean residual sum of squares is obtained by dividing the residual sum of squares by $(r - 1)(c - 1)$. The analysis of the variance test statistic is generated by dividing the mean row sum of squares by the mean residual sum of squares. This value is compared with the tabulated value, $F_{r-1,(r-1)(c-1),\alpha}$, where α is the tolerated value of a Type I error. If the value of the test statistic exceeds the tabulated value, the null hypothesis of no significant difference in the effects of row factor levels is rejected.

The aforementioned procedure can extend to other applications. For example, a two-way classification with more than one observation per cell permits testing the presence of interaction effects in addition to the separate effects of the row factor and the column factor. Interaction is the difference between the combined effect of the row and column factors and the sum of the separate effects of the row and column factors. Interaction partially accounts for the variation not explainable in terms of the separate effects of the row and column factors. Consider a two-way classification with m observations per cell. Let X_{ijk} represent the kth observation in the ith row and jth column; $i = 1, \ldots, r; j = 1, \ldots, c; k = 1, \ldots, m$. Let $X_{ij}, X_i,$ and X_j denote the cell, row and column means, respectively. Accordingly,

$$\overline{X}_{ij.} = \frac{\sum_{k=1}^{m}X_{ijk}}{m} \tag{30.20}$$

$$\overline{X}_{i..} = \frac{\sum_{j=1}^{c}\sum_{k=1}^{m}X_{ijk}}{mc} \tag{30.21}$$

TABLE 30.2
Sum of Squares Data

Group 1	Group 2	Group 3	Group 4	Group 5
3	5	7	6	4
2	8	8	8	9
4	8	6	7	5

$$\bar{X}_{.j.} = \frac{\sum_{i=1}^{r}\sum_{k=1}^{m} X_{ijk}}{mr} \tag{30.22}$$

The total sum of squares can be portioned in a manner similar to that provided in Equation 30.6 to Equation 30.7 and Equation 30.16 to Equation 30.19. Details are provided in Problem ANV 8.

PROBLEMS AND SOLUTIONS

ANV 1: ANOVA

Define ANOVA.

Solution

The acronym ANOVA represents Analysis of Variance.

ANV 2: Sum of Squares Procedure

Consider five groups of academic environmental scientists with three sets of data that represent the annual number of proposals submitted to the USEPA over the past three years (see Table 30.2).
Calculate the following:

1. The among group of squares
2. The residual sum of squares
3. Total sum of squares

Solution

The data in Table 30.2 are first rewritten in Table 30.3 in accordance with the procedure outlined in the Introduction:

$$\sum_{j=1}^{3}\sum_{i=1}^{5}(X_{ij})^2 = 3^2 + 2^2 + 4^2 + \ldots 9^2 + 5^2$$

$$= 602$$

$$n = 5$$

TABLE 30.3
Sum of Squares Calculation

Group 1	Group 2	Group 3	Group 4	Group 5	
3	5	7	6	4	
2	8	8	8	9	
4	8	6	7	5	
$T_{1.} = 9$	$T_{2.} = 21$	$T_{3.} = 21$	$T_{4.} = 21$	$T_{5.} = 18$	$T_{..} = 90$
$M_1 = 3$	$M_2 = 7$	$M_3 = 7$	$M_4 = 7$	$M_5 = 6$	$\overline{T} = \overline{X} = 6$

Note: M and T are the mean and total value, respectively.

1. The among group of squares GSS is (Equation 30.6):

$$GSS = \sum_{i=1}^{5} (T_{i.}^2 / m_i) - (T_{..}^2 / n)$$

$$= \frac{(9)^2 + (21)^2 + (21)^2 + (21)^2 + (18)^2}{3} - \frac{(90)^2}{15}$$

$$= \frac{1728}{3} - \frac{8100}{15}$$

$$= 36$$

2. The residual sum of squares RSS is (Equation 30.7):

$$RSS = \sum_{j=1}^{3} \sum_{i=1}^{5} (X_{ij})^2 - \sum_{i=1}^{k} (T_{i.}^2 / n)$$

$$= 602 - \frac{(9)^2 + (21)^2 + (21)^2 + (21)^2 + (18)^2}{3}$$

$$= 602 - 576$$

$$= 26$$

3. The total sum of squares TSS is

$$TSS = \sum_{j=1}^{3} \sum_{i=1}^{5} X_{ij} - \sum \frac{T_{..}^2}{n}$$

$$= 602 - 540$$

$$= 62$$

Alternatively,

$$TSS = GSS + RSS$$

$$= 36 + 26$$

$$= 62$$

The foregoing analysis can now be extended to include calculations for mean among the group sum of squares, the mean residual sum of squares, and the analysis of variance test statistic.

The mean among group sum of squares (MGSS) is (Equation 30.8)

$$MGSS = \frac{GSS}{k-1}; \quad k = 1$$

$$= \frac{36}{5-1}$$

$$= 9.0$$

The mean residual sum of squares, MRSS, is (Equation 30.9)

$$MRSS = \frac{RSS}{n-k}; \quad n = 15$$

$$= \frac{26}{15-5}$$

$$= 2.6$$

The analysis of variance test statistic, ANVT, is

$$ANVT = \frac{MGSS}{MRSS}$$

$$= \frac{9.0}{2.6}$$

$$= 3.46$$

The ANVT can then be compared to the appropriate tabulated value of F. This is demonstrated in Problem ANV 3.

ANV 3: Comparison of Effect of Food Supplements on Milk Yield

In an experiment testing the effectiveness of different food supplements on milk yield, a sample of 30 cows was divided into four groups. Seven cows were assigned at random to group 1, eight to group 2, eight to group 3, and seven to group 4. The cows in group 1, the control group, received no food supplement, whereas those in group 2 to group 4 received food supplements A, B, and C, respectively. Table 30.4 shows the mean daily milk yield recorded over a 1-month period for each of the cows.

Test the hypothesis that there is no significant difference among the food supplements with respect to their effect on milk yield. Use 0.01 as the tolerated probability of a Type I error.

TABLE 30.4
Data for Effect of Food Supplements
on Yield

Group 1	Group 2	Group 3	Group 4
21.6	22.1	26.0	29.8
16.0	17.6	17.1	30.9
16.0	25.1	25.9	29.9
20.4	25.6	21.5	24.9
22.2	21.6	21.0	23.0
13.9	23.5	28.1	30.8
15.4	24.7	25.5	27.1
N/A	24.1	18.7	N/A

Solution

First obtain the group totals, grand total, and the sum of squares of the observations:

$$T_{1.} = \text{total of group 1}$$
$$= 21.6 + 16.0 + 16.0 + 20.4 + 22.2 + 13.9 + 15.4$$
$$= 125.5$$

$$T_{2.} = \text{total of group 2}$$
$$= 22.1 + 17.6 + 25.1 + 25.6 + 21.6 + 23.5 + 24.7 + 24.1$$
$$= 184.3$$

$$T_{3.} = \text{total of group 3}$$
$$= 26.0 + 17.1 + 25.9 + 21.5 + 21.0 + 28.1 + 25.5 + 18.7$$
$$= 183.8$$

$$T_{4.} = \text{total of group 4}$$
$$= 29.80 + 30.9 + 29.9 + 24.9 + 23.0 + 30.8 + 27.1$$
$$= 196.4$$

$$T_{..} = \text{grand total}$$
$$= 125.5 + 184.3 + 183.8 + 196.4$$
$$= 690.0$$

$$\sum\sum X_{ij}^2 = \text{sum of squares of the observations}$$
$$= (21.6)^2 + (16.0)^2 + \ldots + (27.1)^2$$
$$= 16{,}512.1$$

Compute the among group sum of squares. Employ Equation 30.6:

$$GSS = \sum_{i=1}^{k} (T_{i.}^2 / m_i) - (T_{..}^2 / n)$$

$$= (125.5)^2 / 7 + (184.3)^2 / 8 + (183.8)^2 / 8 + (196.4)^2 / 7 - (690.0)^2 / 30$$

$$= 16,229.1 - 15,870$$

$$= 359.1$$

Compute the residual sum of squares:

$$RSS = \sum_{j=1}^{m_i} \sum_{i=1}^{k} (X_{ij})^2 - \sum_{i=1}^{k} (T_{i.}^2 / m_i)$$

$$= 16,512.1 - 16,229.1$$

$$= 283.0$$

Compute the total sum of squares:

$$TSS = \sum_{j=1}^{8} \sum_{i=1}^{4} X_{ij} - \sum \frac{T_{..}^2}{n}$$

$$= 16,512.1 - 15,870$$

$$= 642.1$$

Compute the mean among group sum of squares:

$$MGSS = (GSS) / (k-1)$$

$$= 359.1 / (4-1)$$

$$= 119.7$$

Compute the mean residual sum of squares:

$$MRSS = (RSS) / (n-k)$$

$$= 283.0 / (30-4)$$

$$= 10.9$$

Finally, obtain the value of the analysis of variance test statistic:

$$ANVT = \frac{MGSS}{MRSS}$$

$$= 119.7 / 10.9$$

$$= 11.0$$

Compare the preceding result with the tabulated value, $F_{k-1,\ N-k;\alpha}$, where α is the tolerated probability of a Type I error. Because

$$\alpha = 0.01$$

$$F_{3,26;0.01} = 4.64 \text{ (linear interpolation)}$$

Because the value of the test statistic, 11.0, exceeds 4.64, the hypothesis of no significant difference among the food supplements with respect to their effect on milk yield is rejected; i.e., there is a difference. Alternatively, because the value of the test statistic exceeds $F_{k-1,n-k,\alpha}$, one would reject the hypothesis of no difference among the mean milk yields of the four populations from which the four groups were chosen.

ANV 4: Variable Type I Error

With reference to Problem ANV 3, test the same hypothesis if the tolerated probability of a Type I error is 0.05.

Solution

For $\alpha = 0.05$,

$$F_{3,26;0.05} = 2.98 \text{ (linear interpolation)}$$

Because $11.0 > 2.98$, one must once again reject the hypothesis of no significant difference.

ANV 5: ANOVA Comparison

Comment on the results of Problem ANV 3 and Problem ANV 4.

Solution

The results of both problems clearly indicate that a higher Type I error can be tolerated before the hypothesis of no significant difference is accepted.

ANV 6: Effect of Rubber Mix and Curing Time on Tensile Strength

Table 30.5 shows the results of an investigation of the tensile strength of four different mixes of rubber at five different curing times.

Test the hypothesis that there is no significant difference among curing times in their effect on tensile strength of rubber produced.

Solution

This problem involves a study of variation attributable to two factors with a two-way classification in which rc observations are classified in r rows and c cells, with one observation in each cell.

In this problem, the row factor is curing time in minutes and the five levels are 15, 30, 45, 60, and 90. The column factor is rubber mix by type. The levels are the different rubber mixes and the four levels are Ameripol SN, Hevea Y, Hevea B, and Hevea TB. Note that there are five rows and four columns.

Compute the row sum of squares using Equation 30.16,

TABLE 30.5
Tensile Strength (lb/in²) of Rubber Mixes at Different Curing Times

Curing Time (min)	Rubber Mix				Total
	Ameripol SN	Hevea Y	Hevea B	Hevea TB	
15	2840	2540	2900	2490	10770
30	3550	3150	3970	3410	14080
45	3440	3850	3810	3210	14310
60	3200	3450	4000	3670	14320
90	3210	3600	3730	3620	14160
Total	16240	16590	18410	16400	67640

$$RSS2 = c \sum (\overline{X}_i - \overline{X})^2$$

$$= \frac{\sum T_{i.}^2}{c} - \frac{T_{..}^2}{rc}$$

$$= \frac{[(10770)^2 + (14080)^2 + (14310)^2 + (14320)^2 + (14160)^2]}{4} - \frac{(67640)^2}{20}$$

$$= 231,145,850 - 228,758,480$$

$$= 2,387,370$$

Compute the column sum of squares using Equation 30.17.

$$CSS2 = r \sum (\overline{X}_{ij} - \overline{X})^2$$

$$= \frac{\sum T_{.j}^2}{r} - \frac{T_{..}^2}{rc}$$

$$= \frac{[(16240)^2 + (16590)^2 + (18410)^2 + (16400)^2]}{5} - \frac{(67640)^2}{20}$$

$$= 229,370,760 - 228,758,480$$

$$= 612,280$$

Compute the total sum of squares using Equation 30.18:

$$TSS2 = \sum \sum (X_{ij} - \overline{X})^2$$

$$= \sum \sum X_{ij}^2 - \frac{T_{..}^2}{rc}$$

$$= 232,410,400 - \frac{(67640)^2}{20}$$

$$= 232,410,400 - 228,758,480$$

$$= 3,651,920$$

Compute the residual sum of squares using Equation 30.19:

$$RRSS2 = \sum\sum (X_{ij} - \overline{X}_{i.} - \overline{X}_{.j} + \overline{X})^2$$

$$= \sum\sum X_{ij}^2 - \frac{\sum T_{i.}^2}{c} - \frac{\sum T_{.j}^2}{r} + \frac{T_{..}^2}{rc}$$

$$= 232,410,400 - 231,145,850 - 229,370,760 + 228,758,480$$

$$= 652,270$$

Obtain the mean row sum of squares (see Introduction):

$$MRSS2 = (RSS2)/(r-1)$$

$$= 2,387,370/(5-1)$$

$$= 596,842.5$$

Obtain the mean column sum of squares:

$$MCSS2 = (CSS2)/(c-1)$$

$$= 312,280/(5-1)$$

$$= 204,903.33$$

Obtain the mean residual sum of squares:

$$MRRSS2 = (RRSS2)/(r-1)(c-1)$$

$$= 652,270/(5-1)(4-1)$$

$$= 54,355.83$$

Now evaluate the analysis of variance test statistic for testing the hypothesis of no significant difference among row factor levels in their effect on tensile strength. For this test,

$$\text{Analysis of variance test statistic} = \frac{\text{Mean row sum of squares}}{\text{Mean residual sum of squares}}$$

$$ANVT = (596,842.5)/(54,355.83)$$

$$= 10.98$$

Compare the value obtained with the tabulated value, $F_{r-1,(r-1)(c-1);\alpha}$, where α is the tolerated probability of a Type I error ($\alpha = 0.01$). From Table A.4A (Appendix A),

$$F_{r-1,(r-1)(c-1);\alpha} = F_{4,12;0.01} = 5.41$$

Because the value of the analysis of variance test statistic, 10.98, exceeds 5.91, one would reject the hypothesis of no significant difference in the effects of curing times on tensile strength and conclude that tensile strength differs significantly at the various levels of curing time.

ANV 7: EFFECT OF RUBBER MIXES

Refer to Problem ANV 6. Test the hypothesis that there is no significant difference among rubber mixes in their effect on tensile strength of rubber produced. Once again, use 0.01 as the tolerated probability of a Type I error.

Solution

Evaluate the analysis of variance test statistic for testing the hypothesis of no significant difference among the column factor levels in their effect on tensile strength:

$$\text{Analysis of variance test statistic} = \frac{\text{Mean row sum of squares}}{\text{Mean residual sum of squares}}$$

$$ANVT = \frac{MGSS2}{MRRSS2}$$

$$= (204,903.33)/(54,355.83)$$

$$= 3.75$$

Compare the value obtained with the tabulated value, $F_{c-1,(r-1)(c-1),\alpha}$, where α is the tolerated probability of a Type I error ($\alpha = 0.01$):

$$F_{rc1,(r-1)(c-1);\alpha} = F_{3,12;0.01} = 5.95$$

Because the value of the analysis of variance test statistic, 3.75, does not exceed 5.95, one would accept the hypothesis of no significant difference in the effect of rubber mix types on tensile strength and conclude that the evidence is not sufficient to conclude that there is a significant difference among rubber mix types in their effect on tensile strength.

ANV 8: INTERACTION EFFECTS

Develop an ANOVA procedure to include interaction effects.

Solution

A two-way classification with more than one observation per cell permits testing the presence of interaction effects in addition to the separate effects of the row factor and the column factor. Interaction is the difference between the combined effect of the row and column factors and the sum of the separate effects of the row and column factors. Interaction partially accounts for the variation not explainable in terms of the separate effects of the row and column factors.

Consider a two-way classification with m observations per cell. Let X_{ijk} denote the kth observation in the ith row and jth column; $i = 1, \ldots, r; j = 1, \ldots, c; k = 1, \ldots, m$. Let X_{ij}, X_i, and X_j. denote the cell, row, and column means, respectively. Accordingly,

$$\overline{X}_{ij.} = \frac{\sum_{k=1}^{m} X_{ijk}}{m} \tag{30.23}$$

$$\overline{X}_{i..} = \frac{\sum_{j=1}^{c}\sum_{k=1}^{m} X_{ijk}}{mc} \tag{30.24}$$

$$\overline{X}_{.j.} = \frac{\sum_{i=1}^{r}\sum_{k=1}^{m} X_{ijk}}{mr} \tag{30.25}$$

The total sum of squares can be partitioned as follows:

$$\sum\sum\sum(X_{ijk} - \overline{X})^2 = mc\sum(\overline{X}_{i..} - \overline{X})^2 + mr\sum(\overline{X}_{.j.} - \overline{X})$$
$$+ m\sum\sum(\overline{X}_{ij.} - \overline{X}_{i..} - \overline{X}_{.j.} + \overline{X})^2 + \sum\sum\sum(X_{ijk} - \overline{X}_{ij.})^2 \tag{30.26}$$

The four components into which the total sum of squares has been partitioned are designated respectively as the row sum of squares, the column sum of squares, the interaction sum of squares (ISS) and the residual sum of squares. The way in which the interaction sum of squares measures the variation due to interaction can be revealed by noting that the interaction sum of squares can be written as

$$\text{Interaction sum of squares} = \sum\sum\sum[(\overline{X}_{ij.} - \overline{X}) - (\overline{X}_{i..} - \overline{X}) - (\overline{X}_{.j.} - \overline{X})]^2 \tag{30.27}$$

An application of this procedure — including alternative calculation methods — is given in Problem ANV 9.

ANV 9: Diet and Drug Effects on Stimulus Response Time

In a study of the effects of three different diets and four different drugs on the time of response to a certain stimulus, 60 experimental animals were assigned at random in groups of 5 to each of the 12 combinations of drug and diet. Table 30.6 shows the observed response time in seconds for each of the animals.

Test the hypothesis that diets I, II, and III do not differ significantly in their effects on stimulus response time. Use 0.01 as the tolerated probability of a Type I error.

Solution

The row factor is diet. The three levels of the row factor are diet I, diet II, and diet III. The column factor is drug. The four levels of the column factor are drug A, drug B, drug C, and drug D.

TABLE 30.6
Response Time in Seconds to Stimulus after Administration of Drugs to Animals on Different Diets

Diet	Drug A	B	C	D	Total
I	6	10	13	16	
	10	8	14	5	
	10	11	10	8	
	16	10	18	11	
	17	16	11	14	234
II	11	11	13	11	
	14	13	11	13	
	16	6	11	16	
	15	10	9	15	
	13	12	10	12	242
III	8	18	16	11	
	8	12	17	8	
	7	14	18	16	
	11	17	18	6	
	13	11	19	10	258
Total	175	179	208	172	734

First identify the possible interaction effect. The interaction effect for diet I and drug A is the difference between the combined effect of diet I and drug A and the sum of the separate effects of diet I and drug A on stimulus response time. The interaction effect in the case of each of the other 11 combinations of diet and drug is defined similarly.

Compute the row sum of squares. See Equation 30.26:

$$\text{Row sum of squares} = mc \sum (\overline{X}_{i..} - \overline{X})^2$$

This may be written as

$$RSS3 = \frac{\sum T_{i..}^2}{mc} - \frac{T_{...}^2}{mrc}$$

$$= \frac{[(234)^2 + (242)^2 + (258)^2]}{20} - \frac{(734)^2}{60}$$

$$= 8994.20 - 8979.27$$

$$= 14.93$$

Compute the column sum of squares. See Equation 30.26:

$$\text{Column sum of squares} = mr \sum (\overline{X}_{.j.} - \overline{X})^2$$

This may be written as

$$CSS3 = \frac{\sum T_{.j.}^2}{mr} - \frac{T_{...}^2}{mrc}$$

$$= \frac{[(175)^2 + (179)^2 + (208)^2 + (172)^2]}{15} - \frac{(734)^2}{60}$$

$$= 9034.27 - 8979.27$$

$$= 55.00$$

Compute the interaction sum of squares by employing Equation 30.15.

$$\text{Interaction sum of squares} = m \sum \sum [(\overline{X}_{ij.} - \overline{X}) - (\overline{X}_{i..} - \overline{X}) - (\overline{X}_{.j.} - \overline{X})]^2$$

This may be written as

$$ISS3 = \frac{\sum \sum T_{ij.}^2}{m} - \frac{\sum T_{i..}^2}{mc} - \frac{\sum T_{.j.}^2}{mr} + \frac{T_{...}^2}{mrc}$$

$$= \frac{[(59)^2 + (55)^2 + (66)^2 + (54)^2 + (69)^2 + (52)^2 + (54)^2 + (67)^2 + (47)^2 + (72)^2 + (88)^2 + (51)^2]}{5}$$

$$- 8994.20 - 9034.27 - 8979.27$$

$$= 9027.20 - 8994.20 - 9034.27 + 8979.27$$

$$= 228.00$$

Compute the residual sum of squares. See Equation 30.26.

$$\text{Residual sum of squares} = \sum \sum \sum (X_{ijk} - \overline{X}_{ij.})^2$$

This can be rewritten as

$$RRSS3 = \sum \sum \sum X_{ijk}^2 - \frac{\sum \sum T_{ij.}^2}{m}$$

$$= 9710 - 9277.20$$

$$= 432.80$$

Compute the total sum of squares. See Equation 30.26.

$$\text{Total sum of squares} = \sum\sum\sum (X_{ijk} - \overline{X})^2$$

This can also be written as

$$TSS3 = \sum\sum\sum X_{ijk}^2 - \frac{\sum\sum T_{ij.}^2}{m}$$

$$= 9710 - 8979.27$$

$$= 730.73$$

Obtain the mean row, the mean column, the mean interaction, and the mean residual sum of squares using the procedure given as follows:

$$\text{Mean row sum of squares} = (\text{Row sum of squares}) / (r-1)$$

$$MRSS3 = RSS3 / (r-1)$$

$$= 14.93 / (3-1)$$

$$= 7.47$$

$$\text{Mean column sum of squares} = (\text{Column sum of squares}) / (c-1)$$

$$MCSS3 = CSS3 / (c-1)$$

$$= 55.00 / (4-1)$$

$$= 18.33$$

$$\text{Mean interaction sum of squares} = (\text{Interaction sum of squares}) / (c-1)(r-1)$$

$$MISS = ISS3 / (c-1)(r-1)$$

$$= 228.00 / (4-1)(3-1)$$

$$= 38.00$$

$$\text{Mean residual sum of squares} = (\text{Residual sum of squares}) / rc(m-1)$$

$$MRSS3 = RRSS3 / rc(m-1)$$

$$= 432.80 / (5-1)(3)(4)$$

$$= 9.02$$

Evaluate the analysis of variance test statistic for testing the hypothesis of no significant difference among the row factor levels in their effect on stimulus response time using the following equation:

$$\text{Analysis of variance test statistic} = \frac{\text{Mean row of squares}}{\text{Mean residual sum of squares}}$$

$$ANVT = MRSS3 \,/\, MRRSS3$$

$$= (7.4) \,/\, (9.02)$$

$$= 0.83$$

Compare the value obtained with the tabulated value $F_{r-1,rc(m-1),\alpha}$, where α is the tolerated probability of a Type I error ($\alpha = 0.01$):

$$F_{r-1,rc(m-1),\alpha} = F_{2,48,0.01} = 5.10 \text{ (linear interpolation)}$$

Because the observed value of the analysis of variance test statistic does not exceed 5.10, one would fail to reject the null hypothesis of no significant difference among the row factor levels in their effect on stimulus response time and conclude that stimulus response times for diet I, diet II, and diet III do not differ significantly. Alternatively, if the value of the test statistic exceeds 5.10, one would reject the null hypothesis of no significant difference among the row factor level effects.

ANV 10: Drug Test

Refer to Problem ANV 9. Test the hypothesis that drugs A, B, C, and D do not differ significantly in their effect on stimulus response time. Use 0.01 as the tolerated probability of a Type I error.

Solution

Evaluate the analysis of variance test statistic for testing the hypothesis of no significant difference among the column factor levels in their effect on stimulus response time:

$$\text{Analysis of variance test statistic} = \frac{\text{Mean column sum of squares}}{\text{Mean residual sum of squares}}$$

$$= (18.33) \,/\, (9.02)$$

$$= 2.03$$

Compare the value obtained with the tabulated value $F_{c-1,rc(m-1),\alpha}$, where α is the tolerated probability of a Type I error ($\alpha = 0.01$):

$$F_{c-1,rc(m-1),\alpha} = F_{3,48,0.01} = 4.24 \text{ (linear interpolation)}$$

Because the value of the analysis of variance test statistic again does not exceed 4.24, one would fail to reject the null hypothesis of no significant difference among column factor levels in their effect on stimulus response time and conclude that stimulus response times for drug A, drug B, drug C, and drug D do not differ significantly.

ANV 11: No Interaction Effect

Refer to Problem ANV 9. Determine if there is no significant diet–drug interaction effect on stimulus response time. Use 0.01 as the tolerated probability of a Type I error.

Solution

One needs to evaluate the analysis of variance test statistic for testing the absence of interaction of row and column factor level effects on stimulus response time:

$$\text{Analysis of variance test statistic} = \frac{\text{Mean interaction sum of squares}}{\text{Mean residual sum of squares}}$$

$$= (38.00)/(9.02)$$

$$= 4.21$$

Compare the value obtained with the tabulated value $F_{(c-1)(r-1),rc(m-1),\alpha}$, where α is the tolerated probability of a Type I error ($\alpha = 0.01$):

$$F_{(c-1)(r-1),rc(m-1),\alpha} = F_{6,48,0.01} = 3.22 \text{ (linear interpolation)}$$

Because the value of the analysis of variance test statistic exceeds 3.22, one would reject the null hypothesis of no interaction of row and column factor level effects and conclude that stimulus response time differs significantly for various combinations of diet and drug.

ANV 12: Level of Significance Change

Refer to Problem ANV 9 through Problem ANV 11. Determine what effect changing the level of significance from 0.01 to 0.05 has on the final results.

Solution

For Problem ANV 9, a new determination for $F_{2,48,0.05}$ must be obtained from Table A.4B (Appendix A):

$$F_{2,48,0.05} = 3.19 \text{ (linear interpolation)}$$

Because

$$3.19 > 0.83$$

the result remains the same.

For Problem ANV 10, a new determination for $F_{3,48,0.05}$ must be obtained from Table A.4B (Appendix A):

$$F_{3,48,0.05} = 2.80 \text{ (linear interpolation)}$$

Because

$$2.80 > 2.03$$

the result also remains the same.

For Problem ANV 11, a new determination for $F_{6,48,0.05}$ must be obtained from Table A.4B (Appendix A):

$$F_{6,48,0.05} = 2.30 \text{ (linear interpolation)}$$

Because

$$2.30 < 4.21$$

the null hypothesis of interaction is again rejected.

Appendix A

Appendix A

TABLE A.1A
Standard Normal Cumulative Probability in Right-Hand Tail (Area Under Curve for Specified Values of z_0)

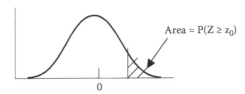

Area $= P(Z \geq z_0)$

The Standard Normal Distribution

z	0.00	0.01	0.02	0.03	0.04	0.05	0.06	0.07	0.08	0.09
0.0	0.500	0.496	0.492	0.488	0.484	0.480	0.476	0.472	0.468	0.464
0.1	0.460	0.456	0.452	0.448	0.444	0.440	0.436	0.433	0.429	0.425
0.2	0.421	0.417	0.413	0.409	0.405	0.401	0.397	0.394	0.390	0.386
0.3	0.382	0.378	0.374	0.371	0.367	0.363	0.359	0.356	0.352	0.348
0.4	0.345	0.341	0.337	0.334	0.330	0.326	0.323	0.319	0.316	0.312
0.5	0.309	0.305	0.302	0.298	0.295	0.291	0.288	0.284	0.281	0.278
0.6	0.274	0.271	0.268	0.264	0.261	0.258	0.255	0.251	0.248	0.245
0.7	0.242	0.239	0.236	0.233	0.230	0.227	0.224	0.221	0.218	0.215
0.8	0.212	0.209	0.206	0.203	0.200	0.198	0.195	0.192	0.189	0.187
0.9	0.184	0.181	0.179	0.176	0.174	0.171	0.169	0.166	0.164	0.161
1.0	0.159	0.156	0.154	0.152	0.149	0.147	0.145	0.142	0.140	0.138
1.1	0.136	0.133	0.131	0.129	0.127	0.125	0.123	0.121	0.119	0.117
1.2	0.115	0.113	0.111	0.109	0.107	0.106	0.104	0.102	0.100	0.099
1.3	0.097	0.095	0.093	0.092	0.090	0.089	0.087	0.085	0.084	0.082
1.4	0.081	0.079	0.078	0.076	0.075	0.074	0.072	0.071	0.069	0.068
1.5	0.067	0.066	0.064	0.063	0.062	0.061	0.059	0.058	0.057	0.056
1.6	0.055	0.054	0.053	0.052	0.051	0.049	0.048	0.047	0.046	0.046
1.7	0.045	0.044	0.043	0.042	0.041	0.040	0.039	0.038	0.038	0.037
1.8	0.036	0.035	0.034	0.034	0.033	0.032	0.031	0.031	0.030	0.029
1.9	0.029	0.028	0.027	0.027	0.026	0.026	0.025	0.024	0.024	0.023
2.0	0.023	0.022	0.022	0.021	0.021	0.020	0.020	0.019	0.019	0.018
2.1	0.018	0.017	0.017	0.017	0.016	0.016	0.015	0.015	0.015	0.014
2.2	0.014	0.014	0.013	0.013	0.013	0.012	0.012	0.012	0.011	0.011
2.3	0.011	0.010	0.010	0.010	0.010	0.009	0.009	0.009	0.009	0.008
2.4	0.008	0.008	0.008	0.008	0.007	0.007	0.007	0.007	0.007	0.006
2.5	0.006	0.006	0.006	0.006	0.006	0.005	0.005	0.005	0.005	0.005
2.6	0.005	0.005	0.005	0.005	0.005	0.004	0.004	0.004	0.004	0.004
2.7	0.003	0.003	0.003	0.003	0.003	0.003	0.003	0.003	0.003	0.003
2.8	0.003	0.002	0.002	0.002	0.002	0.002	0.002	0.002	0.002	0.002
2.9	0.002	0.002	0.002	0.002	0.002	0.002	0.002	0.002	0.002	0.002

Adapted from: http://www.statsoft.com/textbook/sttable.html

TABLE A.1.B
Areas Under a Standard Normal Curve Between 0 and z

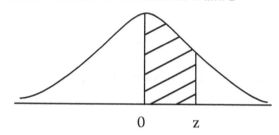

The Standard Normal Distribution

z	0.00	0.01	0.02	0.03	0.04	0.05	0.06	0.07	0.08	0.09
0.0	0.0000	0.0040	0.0080	0.0120	0.0160	0.0199	0.0239	0.0279	0.0319	0.0359
0.1	0.0398	0.0438	0.0478	0.0517	0.0557	0.0596	0.0636	0.0675	0.0714	0.0753
0.2	0.0793	0.0832	0.0871	0.0910	0.0948	0.0987	0.1026	0.1064	0.1103	0.1141
0.3	0.1179	0.1217	0.1255	0.1293	0.1331	0.1368	0.1406	0.1443	0.1480	0.1517
0.4	0.1554	0.1591	0.1628	0.1664	0.1700	0.1736	0.1772	0.1808	0.1844	0.1879
0.5	0.1915	0.1950	0.1985	0.2019	0.2054	0.2088	0.2123	0.2157	0.2190	0.2224
0.6	0.2257	0.2291	0.2324	0.2357	0.2389	0.2422	0.2454	0.2486	0.2517	0.2549
0.7	0.2580	0.2611	0.2642	0.2673	0.2704	0.2734	0.2764	0.2794	0.2823	0.2852
0.8	0.2881	0.2910	0.2939	0.2967	0.2995	0.3023	0.3051	0.3078	0.3106	0.3133
0.9	0.3159	0.3186	0.3212	0.3238	0.3264	0.3289	0.3315	0.3340	0.3365	0.3389
1.0	0.3413	0.3438	0.3461	0.3485	0.3508	0.3531	0.3554	0.3577	0.3599	0.3621
1.1	0.3643	0.3665	0.3686	0.3708	0.3729	0.3749	0.3770	0.3790	0.3810	0.3830
1.2	0.3849	0.3869	0.3888	0.3907	0.3925	0.3944	0.3962	0.3980	0.3997	0.4015
1.3	0.4032	0.4049	0.4066	0.4082	0.4099	0.4115	0.4131	0.4147	0.4162	0.4177
1.4	0.4192	0.4207	0.4222	0.4236	0.4251	0.4265	0.4279	0.4292	0.4306	0.4319
1.5	0.4332	0.4345	0.4357	0.4370	0.4382	0.4394	0.4406	0.4418	0.4429	0.4441
1.6	0.4452	0.4463	0.4474	0.4484	0.4495	0.4505	0.4515	0.4525	0.4535	0.4545
1.7	0.4554	0.4564	0.4573	0.4582	0.4591	0.4599	0.4608	0.4616	0.4625	0.4633
1.8	0.4641	0.4649	0.4656	0.4664	0.4671	0.4678	0.4686	0.4693	0.4699	0.4706
1.9	0.4713	0.4719	0.4726	0.4732	0.4738	0.4744	0.4750	0.4756	0.4761	0.4767
2.0	0.4772	0.4778	0.4783	0.4788	0.4793	0.4798	0.4803	0.4808	0.4812	0.4817
2.1	0.4821	0.4826	0.4830	0.4834	0.4838	0.4842	0.4846	0.4850	0.4854	0.4857
2.2	0.4861	0.4864	0.4868	0.4871	0.4875	0.4878	0.4881	0.4884	0.4887	0.4890
2.3	0.4893	0.4896	0.4898	0.4901	0.4904	0.4906	0.4909	0.4911	0.4913	0.4916
2.4	0.4918	0.4920	0.4922	0.4925	0.4927	0.4929	0.4931	0.4932	0.4934	0.4936
2.5	0.4938	0.4940	0.4941	0.4943	0.4945	0.4946	0.4948	0.4949	0.4951	0.4952
2.6	0.4953	0.4955	0.4956	0.4957	0.4959	0.4960	0.4961	0.4962	0.4963	0.4964
2.7	0.4965	0.4966	0.4967	0.4968	0.4969	0.4970	0.4971	0.4972	0.4973	0.4974
2.8	0.4974	0.4975	0.4976	0.4977	0.4977	0.4978	0.4979	0.4979	0.4980	0.4981
2.9	0.4981	0.4982	0.4982	0.4983	0.4984	0.4984	0.4985	0.4985	0.4986	0.4986
3.0	0.4987	0.4987	0.4987	0.4988	0.4988	0.4989	0.4989	0.4989	0.4990	0.4990

Adapted from: http://www.statsoft.com/textbook/sttable.html

TABLE A.2
Student's t Distribution

ν	Values of t					ν
	$\alpha = 0.10$	$\alpha = 0.05$	$\alpha = 0.025$	$\alpha = 0.01$	$\alpha = 0.005$	
1	3.077684	6.313752	12.70620	31.82052	63.65674	1
2	1.885618	2.919986	4.30265	6.96456	9.92484	2
3	1.637744	2.353363	3.18245	4.54070	5.84091	3
4	1.533206	2.131847	2.77645	3.74695	4.60409	4
5	1.475884	2.015048	2.57058	3.36493	4.03214	5
6	1.439756	1.943180	2.44691	3.14267	3.70743	6
7	1.414924	1.894579	2.36462	2.99795	3.49948	7
8	1.396815	1.859548	2.30600	2.89646	3.35539	8
9	1.383029	1.833113	2.26216	2.82144	3.24984	9
10	1.372184	1.812461	2.22814	2.76377	3.16927	10
11	1.363430	1.795885	2.20099	2.71808	3.10581	11
12	1.356217	1.782288	2.17881	2.68100	3.05454	12
13	1.350171	1.770933	2.16037	2.65031	3.01228	13
14	1.345030	1.761310	2.14479	2.62449	2.97684	14
15	1.340606	1.753050	2.13145	2.60248	2.94671	15
16	1.336757	1.745884	2.11991	2.58349	2.92078	16
17	1.333379	1.739607	2.10982	2.56693	2.89823	17
18	1.330391	1.734064	2.10092	2.55238	2.87844	18
19	1.327728	1.729133	2.09302	2.53948	2.86093	19
20	1.325341	1.724718	2.08596	2.52798	2.84534	20
22	1.321237	1.717144	2.07387	2.50832	2.81876	22
24	1.317836	1.710882	2.06390	2.49216	2.79694	24
26	1.314972	1.705618	2.05553	2.47863	2.77871	26
28	1.312527	1.701131	2.04841	2.46714	2.76326	28
∞	1.281552	1.644854	1.95996	2.32635	2.57583	∞

Adapted from: http://www.statsoft.com/textbook/sttable.html

TABLE A.3
Chi-Square Distribution

Values of $x^2_{\alpha,v}$

v	$\alpha = 0.995$	$\alpha = 0.990$	$\alpha = 0.975$	$\alpha = 0.950$	$\alpha = 0.050$	$\alpha = 0.025$	$\alpha = 0.010$	$\alpha = 0.005$	v
1	0.00004	0.00016	0.00098	0.00393	3.84146	5.02389	6.63490	7.87944	1
2	0.01003	0.02010	0.05064	0.10259	5.99146	7.37776	9.21034	10.59663	2
3	0.07172	0.11483	0.21580	0.35185	7.81473	9.34840	11.34487	12.83816	3
4	0.20699	0.29711	0.48442	0.71072	9.48773	11.14329	13.27670	14.86026	4
5	0.41174	0.55430	0.83121	1.14548	11.07050	12.83250	15.08627	16.74960	5
6	0.67573	0.87209	1.23734	1.63538	12.59159	14.44938	16.81189	18.54758	6
7	0.98926	1.23904	1.68987	2.16735	14.06714	16.01276	18.47531	20.27774	7
8	1.34441	1.64650	2.17973	2.73264	15.50731	17.53455	20.09024	21.95495	8
9	1.73493	2.08790	2.70039	3.32511	16.91898	19.02277	21.66599	23.58935	9
10	2.15586	2.55821	3.24697	3.94030	18.30704	20.48318	23.20925	25.18818	10
11	2.60322	3.05348	3.81575	4.57481	19.67514	21.92005	24.72497	26.75685	11
12	3.07382	3.57057	4.40379	5.22603	21.02607	23.33666	26.21697	28.29952	12
13	3.56503	4.10692	5.00875	5.89186	22.36203	24.73560	27.68825	29.81947	13
14	4.07467	4.66043	5.62873	6.57063	23.68479	26.11895	29.14124	31.31935	14
15	4.60092	5.22935	6.26214	7.26094	24.99579	27.48839	30.57791	32.80132	15
16	5.14221	5.81221	6.90766	7.96165	26.29623	28.84535	31.99993	34.26719	16
17	5.69722	6.40776	7.56419	8.67176	27.58711	30.19101	33.40866	35.71847	17
18	6.26480	7.01491	8.23075	9.39046	28.86930	31.52638	34.80531	37.15645	18
19	6.84397	7.63273	8.90652	10.11701	30.14353	32.85233	36.19087	38.58226	19
20	7.43384	8.26040	9.59078	10.85081	31.41043	34.16961	37.56623	39.99685	20
22	8.64272	9.54249	10.98232	12.33801	33.92444	36.78071	40.28936	42.79565	22
24	9.88623	10.85636	12.40115	13.84843	36.41503	39.36408	42.97982	45.55851	24
26	11.16024	12.19815	13.84390	15.37916	38.88514	41.92317	45.64168	48.28988	26
28	12.46134	13.56471	15.30786	16.92788	41.33714	44.46079	48.27824	50.99338	28
30	13.78672	14.95346	16.79077	18.49266	43.77297	46.97924	50.89218	53.67196	30

Adapted from: http://www.statsoft.com/textbook/sttable.html

TABLE A.4A
***F* Distribution (a)**

Values of $F_{0.01,\nu_1,\nu_2}$

ν_1 = Degrees of Freedom for Numerator

ν_2	1	2	3	4	5	6	7	8	9	10	15	20	30	40	60	120	∞
1	4052.1	4999.5	5403.4	5624.6	5763.7	5859.0	5928.4	5981.1	6022.5	6055.8	6157.3	6208.7	6260.6	6286.8	6313.0	6339.4	6365.9
2	98.5	99.0	99.17	99.25	99.3	99.33	99.36	99.37	99.39	99.4	99.43	99.45	99.47	99.47	99.48	99.49	99.5
3	34.12	30.82	29.46	28.71	28.24	27.91	27.67	27.49	27.35	27.23	26.87	26.69	26.51	26.41	26.32	26.22	26.13
4	21.2	18.0	16.69	15.98	15.52	15.21	14.98	14.8	14.66	14.55	14.2	14.02	13.84	13.75	13.65	13.56	13.46
5	16.26	13.27	12.06	11.39	10.97	10.67	10.46	10.29	10.16	10.05	9.722	9.553	9.379	9.291	9.202	9.112	9.02
6	13.75	10.93	9.78	9.148	8.746	8.466	8.26	8.102	7.976	7.874	7.559	7.396	7.229	7.143	7.057	6.969	6.88
7	12.25	9.547	8.451	7.847	7.46	7.191	6.993	6.84	6.719	6.62	6.314	6.155	5.992	5.908	5.824	5.737	5.65
8	11.26	8.649	7.591	7.006	6.632	6.371	6.178	6.029	5.911	5.814	5.515	5.359	5.198	5.116	5.032	4.946	4.859
9	10.56	8.022	6.992	6.422	6.057	5.802	5.613	5.467	5.351	5.257	4.962	4.808	4.649	4.567	4.483	4.398	4.311
10	10.04	7.559	6.552	5.994	5.636	5.386	5.2	5.057	4.942	4.849	4.558	4.405	4.247	4.165	4.082	3.996	3.909
11	9.646	7.206	6.217	5.668	5.316	5.069	4.886	4.744	4.632	4.539	4.251	4.099	3.941	3.86	3.776	3.69	3.602
12	9.33	6.927	5.953	5.412	5.064	4.821	4.64	4.499	4.388	4.296	4.01	3.858	3.701	3.619	3.535	3.449	3.361
13	9.074	6.701	5.739	5.205	4.862	4.62	4.441	4.302	4.191	4.1	3.815	3.665	3.507	3.425	3.341	3.255	3.165
14	8.862	6.515	5.564	5.035	4.695	4.456	4.278	4.14	4.03	3.939	3.656	3.505	3.348	3.266	3.181	3.094	3.004
15	8.683	6.359	5.417	4.893	4.556	4.318	4.142	4.004	3.895	3.805	3.522	3.372	3.214	3.132	3.047	2.959	2.868
16	8.531	6.226	5.292	4.773	4.437	4.202	4.026	3.89	3.78	3.691	3.409	3.259	3.101	3.018	2.933	2.845	2.753
17	8.4	6.112	5.185	4.669	4.336	4.102	3.927	3.791	3.682	3.593	3.312	3.162	3.003	2.92	2.835	2.746	2.653
18	8.285	6.013	5.092	4.579	4.248	4.015	3.841	3.705	3.597	3.508	3.227	3.077	2.919	2.835	2.749	2.66	2.566
19	8.185	5.926	5.01	4.5	4.171	3.939	3.765	3.631	3.523	3.434	3.153	3.003	2.844	2.761	2.674	2.584	2.489
20	8.096	5.849	4.938	4.431	4.103	3.871	3.699	3.564	3.457	3.368	3.088	2.938	2.778	2.695	2.608	2.517	2.421
21	8.017	5.78	4.874	4.369	4.042	3.812	3.64	3.506	3.398	3.31	3.03	2.88	2.72	2.636	2.548	2.457	2.36
22	7.945	5.719	4.817	4.313	3.988	3.758	3.587	3.453	3.346	3.258	2.978	2.827	2.667	2.583	2.495	2.403	2.305
23	7.881	5.664	4.765	4.264	3.939	3.71	3.539	3.406	3.299	3.211	2.931	2.781	2.62	2.535	2.447	2.354	2.256
24	7.823	5.614	4.718	4.218	3.895	3.667	3.496	3.363	3.256	3.168	2.889	2.738	2.577	2.492	2.403	2.31	2.211
25	7.77	5.568	4.675	4.177	3.855	3.627	3.457	3.324	3.217	3.129	2.85	2.699	2.538	2.453	2.364	2.27	2.169
26	7.721	5.526	4.637	4.14	3.818	3.591	3.421	3.288	3.182	3.094	2.815	2.664	2.503	2.417	2.327	2.233	2.131
27	7.677	5.488	4.601	4.106	3.785	3.558	3.388	3.256	3.149	3.062	2.783	2.632	2.47	2.384	2.294	2.198	2.097
28	7.636	5.453	4.568	4.074	3.754	3.528	3.358	3.226	3.12	3.032	2.753	2.602	2.44	2.354	2.263	2.167	2.064
29	7.598	5.42	4.538	4.045	3.725	3.499	3.33	3.198	3.092	3.005	2.726	2.574	2.412	2.325	2.234	2.138	2.034
30	7.562	5.39	4.51	4.018	3.699	3.473	3.304	3.173	3.067	2.979	2.7	2.549	2.386	2.299	2.208	2.111	2.006
40	7.314	5.179	4.313	3.828	3.514	3.291	3.124	2.993	2.888	2.801	2.522	2.369	2.203	2.114	2.019	1.917	1.805
60	7.077	4.977	4.126	3.649	3.339	3.119	2.953	2.823	2.718	2.632	2.352	2.198	2.028	1.936	1.836	1.726	1.601
120	6.851	4.787	3.949	3.48	3.174	2.956	2.792	2.663	2.559	2.472	2.192	2.035	1.86	1.763	1.656	1.533	1.381
∞	6.635	4.605	3.782	3.319	3.017	2.802	2.639	2.511	2.407	2.321	2.039	1.878	1.696	1.592	1.473	1.325	1

ν_2 = Degrees of freedom.

Adapted from: http://www.statsoft.com/textbook/sstable.html

TABLE A.4.B
F Distribution (b)

Values of $F_{0.05, v_1, v_2}$

v_1 = Degrees of Freedom →

v_2	1	2	3	4	5	6	7	8	9	10	15	20	30	40	60	120	∞
1	161.4	199.5	215.7	224.6	230.2	234	236.8	238.9	240.5	241.9	245.9	248	250.1	251.1	252.2	253.3	254.3
2	18.51	19	19.16	19.25	19.3	19.33	19.35	19.37	19.38	19.4	19.43	19.45	19.46	19.47	19.48	19.49	19.5
3	10.13	9.552	9.277	9.117	9.014	8.941	8.887	8.845	8.812	8.786	8.703	8.66	8.617	8.594	8.572	8.549	8.526
4	7.709	6.944	6.591	6.388	6.256	6.163	6.094	6.041	5.999	5.964	5.858	5.803	5.746	5.717	5.688	5.658	5.628
5	6.608	5.786	5.41	5.192	5.05	4.95	4.876	4.818	4.773	4.735	4.619	4.558	4.496	4.464	4.431	4.399	4.365
6	5.987	5.143	4.757	4.534	4.387	4.284	4.207	4.147	4.099	4.06	3.938	3.874	3.808	3.774	3.74	3.705	3.669
7	5.591	4.737	4.347	4.12	3.972	3.866	3.787	3.726	3.677	3.637	3.511	3.445	3.376	3.34	3.304	3.267	3.23
8	5.318	4.459	4.066	3.838	3.688	3.581	3.501	3.438	3.388	3.347	3.218	3.15	3.079	3.043	3.005	2.967	2.928
9	5.117	4.257	3.863	3.633	3.482	3.374	3.293	3.23	3.179	3.137	3.006	2.937	2.864	2.826	2.787	2.748	2.707
10	4.965	4.103	3.708	3.478	3.326	3.217	3.136	3.072	3.02	2.978	2.845	2.774	2.7	2.661	2.621	2.58	2.538
11	4.844	3.982	3.587	3.357	3.204	3.095	3.012	2.948	2.896	2.854	2.719	2.646	2.571	2.531	2.49	2.448	2.405
12	4.747	3.885	3.49	3.259	3.106	2.996	2.913	2.849	2.796	2.753	2.617	2.544	2.466	2.426	2.384	2.341	2.296
13	4.667	3.806	3.411	3.179	3.025	2.915	2.832	2.767	2.714	2.671	2.533	2.459	2.38	2.339	2.297	2.252	2.206
14	4.6	3.739	3.344	3.112	2.958	2.848	2.764	2.699	2.646	2.602	2.463	2.388	2.308	2.266	2.223	2.178	2.131
15	4.543	3.682	3.287	3.056	2.901	2.791	2.707	2.641	2.588	2.544	2.403	2.328	2.247	2.204	2.16	2.114	2.066
16	4.494	3.634	3.239	3.007	2.852	2.741	2.657	2.591	2.538	2.494	2.352	2.276	2.194	2.151	2.106	2.059	2.01
17	4.451	3.592	3.197	2.965	2.81	2.699	2.614	2.548	2.494	2.45	2.308	2.23	2.148	2.104	2.058	2.011	1.96
18	4.414	3.555	3.16	2.928	2.773	2.661	2.577	2.51	2.456	2.412	2.269	2.191	2.107	2.063	2.017	1.968	1.917
19	4.381	3.522	3.127	2.895	2.74	2.628	2.544	2.477	2.423	2.378	2.234	2.156	2.071	2.026	1.98	1.93	1.878
20	4.351	3.493	3.098	2.866	2.711	2.599	2.514	2.447	2.393	2.348	2.203	2.124	2.039	1.994	1.946	1.896	1.843
30	4.171	3.316	2.922	2.69	2.534	2.421	2.334	2.266	2.211	2.165	2.015	1.932	1.841	1.792	1.74	1.684	1.622
40	4.085	3.232	2.839	2.606	2.45	2.336	2.249	2.18	2.124	2.077	1.925	1.839	1.744	1.693	1.637	1.577	1.509
60	4.001	3.15	2.758	2.525	2.368	2.254	2.167	2.097	2.04	1.993	1.836	1.748	1.649	1.594	1.534	1.467	1.389
120	3.92	3.072	2.68	2.447	2.29	2.175	2.087	2.016	1.959	1.911	1.751	1.659	1.554	1.495	1.429	1.352	1.254
∞		2.996	2.605	2.372	2.214	2.099	2.01	1.938	1.88	1.831	1.666	1.571	1.459	1.394	1.318	1.221	1

v_2 = Degrees of freedom.

Adapted from: http://www.statsoft.com/textbook/sttable.html

Appendix B

Appendix B

Appendix B
Glossary of Terms

1. **Cumulative distribution function, cdf, of a random variable:** A function F defined for all real numbers such that $F(x) = P(X \leq x)$. $F(x)$ is the cumulative sum of all probabilities assigned to real numbers less than or equal to x.

    Example 1: X has the pdf defined by

    $$f(x) = 0; \quad x < -2$$
    $$= 0.25; \quad -2 \leq x \leq 2$$
    $$= 0; \quad x > 2$$

    See Figure below.

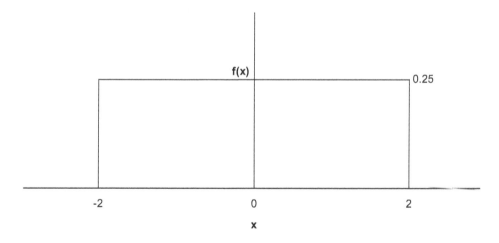

    Then the cdf of X is defined by

    $$F(x) = 0; \quad x < -2$$
    $$= 0.25(x+2); \quad -2 \leq x \leq 2$$
    $$= 1; \quad x > 2$$

Example 2: X has the pdf defined by

$$f(x) = e^{-x}; \quad x > 0$$

Then the cdf of X is defined by

$$F(x) = \int_{-\infty}^{x} f(x)\,dx$$

Therefore,

$$F(x) = 0; \quad x < 0$$

$$= \int_{0}^{x} e^{-x}\,dx = 1 - e^{-x}; \quad x \geq 0$$

2. **Discrete random variable:** A random variable whose range of possible values consists of a finite or countable number of values. The random variables X and Y described earlier are discrete random variables. X has a finite number of values, namely, 0 and 1. Y has an infinite number of values, but they are countable, i.e., there is a first, second, third, and so on.

3. **Event:** A subset of the sample space.

   Example: Let A be the event of at least one head when a coin is tossed twice.

   Then $A = \{HH, TH, HT\}$ if the sample space is $\{HH, TH, HT, TT\}$.

4. **Expected value of a random variable:** The average value of the random variable. If X is the random variable, its expected value is denoted by $E(X)$.

5. **Parameter:** A constant describing the population, e.g., the population mean, the population variance.

6. **Pdf of a continuous random variable:** A function when integrated over an interval gives the probability that the random variable assumes values in the interval. If f is the pdf of a continuous random variable X then

$$\int_{a}^{b} f(x)\,dx = P(a < X < b)$$

Example: For the random variable T above a possible pdf might be defined as follows:

$$f(t) = e^{-t}; \quad t > 0$$

Then,

$$P(1 < T < 3) = \int_{1}^{3} e^{-t}\,dt = e^{-1} - e^{-3} = 0.367 - 0.050 = 0.317$$

7. **Population:** The totality of elements under consideration in a statistical investigation. These elements may be people, objects, or measurements associated with them. Frequently, the population will be the set of possible values of a random variable.

8. **Probability distribution function, pdf, of a discrete random variable:** A function that assigns probability to each of the possible values of the discrete random variable. If f is the pdf of a discrete random variable X then $f(x) = P(X = x)$ is the probability assigned to the value x of the random variable X.

 Example 1: For the random variable X (see 2) defined earlier a possible pdf might be defined as follows:

$$f(x) = 0.01; \quad x = 1$$

$$= 0.99; \quad x = 0$$

 This pdf would assign probability 0.01 to drawing a defective transistor, and 0.99 to drawing a nondefective transistor.

 Example 2: For the random variable Y defined above a possible pdf might be defined as follows:

$$g(y) = (0.001)(0.999)^{y-1}; \quad y = 1, 2, \ldots, n, \ldots 3$$

 Then $P(Y = 1) = 0.0001$, $P(Y = 2) = (0.001)\,(0.999)$, $P(Y = 3) = (0.001)\,(0.999)^2$, etc.

9. **Probability of event A, $P(A)$:** A number about which the relative frequency of event A tends to cluster when the random experiment is repeated indefinitely. This interpretation of $P(A)$ is the relative frequency interpretation. Another interpretation relates $P(A)$ to the degree of belief, measured on a scale from 0 (low) to 1 (high), that the event A occurs.

10. **Random experiment:** An experiment whose outcome is uncertain.

11. **Random variable:** A variable describing the outcome of a random experiment. The values of a random variable are real numbers associated with the elements of the sample space.

 Example 1: Let X denote the result of classifying a transistor drawn at random from a lot as defective or nondefective. Associate 1 with the drawing of a defective and 0 with the drawing of a nondefective.

 Example 2: Let T denote the time to failure of a bus section in an electrostatic precipitator. Then T is a random variable whose values consist of the positive real numbers representing the possible times to failure.

 Example 3: Let Y denote the number of throws of a switch prior to the first failure. Then Y is a random variable with values 1, 2, …, n, ….

 Note that in Example 1 the outcomes of the random experiment had to be assigned a numerical code to define the random variable. In Example 2 and Example 3 the outcome of the random experiment was described numerically, and the numerical values thus introduced became the values of the random variable.

12. **Relative frequency of an event:** The ratio m/n where n is the number of times the random experiment is performed and m is the number of times the event occurs.

 Example: Toss a coin 100 times. Let A be the event of heads on a single toss. Suppose A occurs 40 times. Then $m = 40$, $n = 100$, and the relative frequency of A is 40/100 or 0.40.

13. **Sample:** A subset of the population.
14. **Sample space:** The set of possible outcomes of a random experiment.

> Example: Toss a coin twice. Sample space = {HH, HT, TH, TT} or {0, 1, 2 heads} or {0, 1, 2 tails}. Note that the description of a sample space is not unique.

15. **Set notation for events:**

$$
\begin{array}{ll}
\overline{A} & \text{A does not occur} \\
A \cup B & \text{A or B or both occur} \\
A + B & \text{A or B or both occur (mutually exclusive)} \\
A \cap B & \text{A occurs and B occurs} \\
AB & \text{A occurs and B occurs (independent)} \\
\overline{A}B & \text{A does not occur and B occurs}
\end{array}
$$

16. **Standard deviation of a random variable:** The positive square root of the variance of the random variable. The standard deviation is expressed in the same units as the random variable.
17. **Statistic:** A variable describing the sample, e.g., the sample mean, the sample variance. A statistic is a variable in the sense that its value varies from sample to sample from the same population. A parameter, on the other hand, is constant for any given population.

> Example: Consider a lot of 10,000 transistors each of which is either defective or non-defective. Suppose a sample of 100 transistors is drawn at random from the lot. The proportion of transistors in the lot that are defective is a population parameter. The proportion of transistors in the sample that are defective is a statistic whose value varies from sample to sample drawn from the lot.

18. **Variance of a random variable:** The expected value of the squared difference between a random variable and its mean. If X is a random variable with mean μ, then the variance of X is $E(X - \mu)^2$ and is denoted by σ^2.

SECTION II

The reader is referred to Part I of the Glossary. Many of the terms presented there apply as well to Part II.

SECTION III

1. **Alternative hypothesis:** Possible assumptions, other than the null hypothesis. The alternative hypothesis is usually denoted by H_1. For example, for H_0: $\mu = 20$, H_1 might be $\mu > 20$, or $\mu \neq 20$.
2. **Confidence coefficient:** A measure of confidence, expressed as a percentage between 0 and 100, in the truth of a confidence interval.
3. **Confidence interval:** An interval estimate of an unknown population parameter. For example, on the basis of a sample, the population mean may be estimated to lie between 48 and 52. The interval from 48 to 52 is a confidence interval for the population mean.
4. **Continuous random variable:** A random variable whose range of possible values consists of a noncountable infinitude of values. The random variable T is a continuous

random variable assuming a noncountable infinitude of values represented by the positive real numbers.

5. **Critical region:** Values of the test statistic for which the null hypothesis is rejected.
6. **Experiment:** The process of obtaining an observation. Examples include (1) measuring interplanetary distance, (2) counting the number of students enrolled in a course, (3) tossing a coin, (4) counting the number of throws of a switch prior to the first failure, (5) observing the time to failure of a bus section in an electrostatic precipitator, and (6) observing the date on which a solar eclipse occurs.
7. **Level of significance:** Probability of a Type I error.
8. **Mean of a random variable:** The expected value of the random variable.
9. **Null hypothesis:** The hypothesis being tested. The null hypothesis is usually denoted by H_0.
10. **Point estimate:** A single value estimate of an unknown population parameter. For example, on the basis of the value of the sample mean, the population mean may be estimated to have the value 50. The number 50 is a point estimate of the population mean.
11. **Power of a test:** The probability of rejecting the null hypothesis.
12. **P-value:** Probability of obtaining a value for the test statistic more extreme than the value actually observed.
13. **Statistical hypothesis:** An assumption about one or more random variables, e.g., $\mu = 20$, $\sigma^2 > 5$, X and Y are independent.
14. **Statistical inference:** The process of inferring information from a sample about the population from which the sample is assumed to have been drawn. Probability is used to measure the risk of error in statistical inference.
15. **Test of a statistical hypothesis:** A rule for deciding when to reject the hypothesis. For example, reject H_0: $\mu = 8$ when the sample mean exceeds 8.2.
16. **Test statistic:** The statistic in terms of which the test of the null hypothesis is formulated. For example, in a test that rejects H_0: $\sigma^2 = 4$ when the sample variance s^2 exceeds 4.5, the test statistic is s^2.
17. **Type I error:** Rejecting the null hypothesis when it is true.
18. **Type II error:** Failing to reject the null hypothesis when it is false.
19. **Unbiased estimator:** A statistic whose expected value is equal to the parameter being estimated. For example, the sample mean is an unbiased estimator of the population mean.

Appendix C

Appendix C
Introduction to Design of Experiments

One of the more difficult decisions facing the author during the writing of this text was how to handle/treat the more advanced subject of "design of experiments" (DOE). The material presented in the text (Part I, Part II, and Part III) discussed statistical techniques employed to analyze data obtained from "experiments," with little or no consideration given to the experiment itself. For most individuals desiring an understanding of statistics, the planning and design of experiments is not only outside the control, but also beyond the required background of that individual. It is only on rare occasions that the procedures and details of this topic are mandated. Ultimately, it was decided primarily to provide a very short introduction to DOE because of the enormous depth of the topic. That introduction follows.

The terminology employed in DOE is different to some degree from that employed earlier. Traditionally, experimental variables are usually referred to as factors. The factor may be continuous or discrete. Continuous factors include pollutant concentration, temperature, and drug dosage; discrete factors include year, types of process, time, machine operator, and drug classification. The particular value of the variable is defined as the level of the factor. For example, if one were interested in studying a property (physical or chemical) of a nanoparticle at two different sizes, there are two levels of the factor (in this case, the nanoparticle's size). If a combination of factors is employed, they are called treatments with the result defined as the effect. If the study is to include the effect of the shape of the nanoparticle, these different shapes are called *blocks*. Repeating an experiment at the same conditions is called a *replication*.

In a very general sense, the overall subject of DOE can be thought of as consisting of four steps:

1. Determining the importance of (sets of) observations that can be made to the variable or variables one is interested in
2. The procedure by which the observations will be made, i.e., how the data are gathered
3. The review of the observations, i.e., how the data are treated/examined
4. The analysis of the final results

The reader should also note that obtaining data can be

1. Difficult
2. Time consuming
3. Dangerous
4. Expensive
5. Affected by limited quantities available for experimentation

For these reasons there are obviously significant advantages to expending time on the design of an experiment before the start of the experiment. The variables can have a significant effect on the final results; careful selection can prove invaluable. In summary, the purpose of experimental design is to obtain as much information concerning the response as possible with the minimum amount of experimental work possible.

Index